Buildability in Practice

Buildability in Practice: Shaping a Sustainable and Equitable Built Environment explores the often-overlooked concept of buildability in construction, how easily and efficiently a design can be built. It offers practical strategies and real-world examples to improve project outcomes, reduce waste, and promote sustainability. Readers will learn:

- How rationalisation encourages critical evaluation of design choices to reduce complexity and cost.
- How simplicity theory advocates for straightforward, user-friendly construction solutions.
- How standardisation promotes uniformity in design elements, enhancing speed, quality, and compatibility across projects.
- How system theory integrates all these elements, helping practitioners view buildability as a holistic approach that aligns design, planning, and construction processes.

The authors show that a limited understanding of buildability principles, coupled with the failure to integrate them into the project procurement process, significantly contributes to the suboptimal performance observed in construction projects. They then address this issue by providing clear explanations of buildability principles, practical integration strategies in construction project delivery, and testimonies that demonstrate successful implementation.

This book serves as a comprehensive educational resource for stakeholders, including clients, contractors, and consultants, promoting a shared understanding and fostering collaboration from the early stages of project planning. Buildability is a universal challenge in construction, affecting project efficiency, sustainability, and cost worldwide. With increasing global demand for eco-friendly and efficient building practices, this book offers practical and theoretical insights relevant to professionals, academics, and policymakers across the UK, Europe, North America, and beyond.

Innocent Chigozie Osuizugbo is a research fellow in the Department of Construction Management and Quantity Surveying, Faculty of Engineering and the Built Environment, University of Johannesburg, Doornfontein Campus, Johannesburg, South Africa.

Bankole Osita Awuzie is Professor of Smart and Sustainable Built Environments in the Department of Construction Management and Quantity Surveying, University of Johannesburg, South Africa.

Buildability in Practice

Shaping a Sustainable and Equitable Built Environment

Innocent Chigozie Osuizugbo and Bankole Osita Awuzie

Routledge
Taylor & Francis Group
LONDON AND NEW YORK

Cover credit: Shutterstock

First published 2027
by Routledge
4 Park Square, Milton Park, Abingdon, Oxon OX14 4RN

and by Routledge
605 Third Avenue, New York, NY 10158

Routledge is an imprint of the Taylor & Francis Group, an informa business

British Library Cataloguing-in-Publication Data
A catalogue record for this book is available from the British Library

ISBN: 9781041255758 (hbk)
ISBN: 9781041236627 (pbk)
ISBN: 9781003747284 (ebk)

DOI: 10.1201/9781003747284

Typeset in Times New Roman
by codeMantra

Contents

Figures

Tables

Preface

The built environment is central to human well-being, economic development, and environmental sustainability. Over the past decades, the construction industry has faced increasing complexity, fragmentation, and challenges in delivering projects efficiently, safely, and sustainably. This book, *Buildability in Practice: Shaping a Sustainable and Equitable Built Environment*, is intended to provide a comprehensive exploration of the concept of buildability, its evolution, theoretical underpinnings, practical applications, and future potential in creating sustainable and equitable construction outcomes.

The work begins by tracing the historical evolution of buildability, highlighting how early industrialisation, the separation between designers and builders, and growing concerns about productivity laid the foundation for systematised buildability thinking. Subsequent chapters examine conceptual frameworks, theoretical perspectives, and practical applications, including buildability-led procurement, early contractor involvement, and assessment methodologies. This book also addresses the critical link between buildability and health and safety, stakeholder engagement, education and training, and the integration of sustainable construction practices.

A central aim of this book is to demonstrate that buildability is not merely a technical consideration but a socio-technical capability that spans design, construction, policy, and education. By synthesising historical insights, theoretical frameworks, and contemporary practices, this work seeks to provide guidance to practitioners, policymakers, educators, and researchers striving for improved project outcomes, reduced waste, enhanced safety, and inclusive, sustainable development in the built environment.

I extend my sincere gratitude to the numerous professionals, academic mentors, and institutions whose insights, data, and critical feedback have shaped this book. Their contributions have been invaluable in grounding this work in both scholarly rigor and practical relevance. Special thanks are also due to my family and colleagues for their unwavering support and encouragement throughout the research and writing process.

It is my hope that this book will serve as a meaningful reference for understanding and advancing buildability as a fundamental driver of efficiency, sustainability, and equity in construction practice.

Dr. Innocent Chigozie Osuizugbo
March 2026

1 Introduction

1.1 Background and Context

The built environment, comprising buildings, infrastructure, and interconnected systems, remains at the core of societal development, economic productivity, and human well-being. As cities expand and global construction needs intensify, there is increasing pressure on professionals to deliver projects that are efficient, cost-effective, sustainable, and user-focused. Within this complex and evolving landscape, buildability has emerged as a fundamental concept shaping how construction projects are conceived, designed, procured, executed, and maintained. Buildability refers broadly to the extent to which design and planning decisions facilitate ease of construction, minimise complexity, optimise resource use, and enhance overall project performance. It is concerned with the practical realisation of design intent, ensuring that ideas are not only theoretically sound but also constructible, affordable, durable, sustainable, and responsive to the needs of all stakeholders.

The practice of construction has always required negotiations between design creativity and construction feasibility. Historically, this negotiation relied heavily on the experience of master builders and craftsmen who carried design concepts from drawing boards to physical structures. However, the increasing specialisation of professional roles, growing technological sophistication, and the fragmentation of modern construction supply chains have often widened the gap between design and construction. In many contexts, designers operate in isolation from contractors and end users, leading to designs that are aesthetically appealing but structurally inefficient, difficult to construct, and costly to maintain. Buildability, therefore, has become a critical framework for bridging this gap through early collaboration, systematic consideration of construction factors, rationalisation of design elements, and integration of interdisciplinary knowledge.

Globally, the construction industry faces significant challenges, including cost overruns, delays, quality failures, safety incidents, material wastage, and sustainability pressures. Many of these problems stem from design deficiencies and poor constructability considerations. As a result, governments, researchers, and practitioners have increasingly emphasised the need to embed buildability principles into construction processes. Whether through legislation (such as Singapore's buildable design regulations), procurement reforms (including early contractor

DOI: 10.1201/9781003747284-1

involvement), or innovative technologies (such as Building Information Modelling and modular construction), buildability has become a strategic driver of construction modernisation.

At the same time, contemporary construction practice is situated within powerful global imperatives, sustainable development, climate resilience, resource efficiency, and equity in the built environment. As nations work toward net-zero targets, circular economy adoption, and inclusive urban growth, buildability has evolved beyond simply simplifying construction processes. It now encompasses considerations of environmental performance, lifecycle resource optimisation, maintainability, health and safety, and social value creation. This broadened view recognises that buildings are not merely products of technical activities but are deeply interwoven with ecological systems, user well-being, cultural identities, and long-term societal futures.

In many developing contexts, including African countries, buildability challenges are intensified by skills shortages, informality, infrastructure gaps, procurement inefficiencies, and regulatory inconsistencies. A sustained understanding of buildability, its evolution, application, and future role is therefore necessary not only to enhance project performance but also to shape a sustainable and equitable built environment that meets the needs of diverse populations.

This book, *Buildability in Practice: Shaping a Sustainable and Equitable Built Environment*, responds to this global and contextual need by providing a comprehensive discourse on buildability, from foundational theories to applied practices, educational imperatives, and future pathways for sustainable construction. It integrates established knowledge with emerging perspectives to guide practitioners, educators, researchers, and policymakers towards a deeper understanding of buildability as a transformative force within the construction industry.

1.2 Rationale for the Book

Although research on buildability has expanded over the past four decades, much of the existing literature remains fragmented, context-specific, or narrowly focused. Several handbooks address aspects of constructability or design-for-construction, yet very few provide a holistic, integrated exposition of buildability that links conceptual foundations with practical applications, sustainability considerations, stakeholder roles, and future-oriented technological innovations.

There is also a growing need for literature that critically examines buildability within the context of evolving construction realities, such as modularisation, advanced digital tools, green building standards, net-zero commitments, circular economy initiatives, resilience planning, and the increasing demand for equitable access to built infrastructure. These global shifts demand an updated understanding of buildability that extends beyond traditional considerations of construction ease or material rationalisation.

Additionally, while buildability research has made notable progress in high-income countries, the experiences of developing regions remain underexplored. Many low- and middle-income economies face systemic challenges, weak design processes, limited professional integration, skills deficits, and rapid urbanisation that make buildability improvement crucial for sustainable development. This book aims to contribute to closing that gap by drawing on global lessons while acknowledging context-sensitive practices relevant to emerging economies.

Finally, educational institutions and training bodies have increasingly recognised the need to embed buildability competencies in construction curricula. The industry now expects graduates to possess not only theoretical knowledge but also strong constructability awareness and practical problem-solving skills. Yet, academic content relating to buildability remains inconsistent or underdeveloped in many programmes. This book addresses that gap by dedicating a section to pedagogy, curriculum design, and capacity building for buildability competence.

1.3 Aim of the Book

The primary aim of this book is to provide a comprehensive, integrative, and practice-oriented exposition of buildability that advances understanding, enhances professional competence, and supports sustainable and equitable construction outcomes. It seeks to consolidate fragmented literature and reinterpret buildability through a contemporary lens shaped by sustainability, technological advancement, and social responsibility.

1.4 Objectives of the Book

To achieve the stated aim, this book pursues the following specific objectives:

1. To provide a detailed historical and conceptual foundation of buildability, tracing its evolution, principles, theoretical underpinnings, and relationship to related concepts such as constructability and maintainability.
2. To highlight key factors, challenges, and strategies influencing buildability across project phases, from planning and design to procurement, construction, and maintenance.
3. To explore practical applications of buildability in modern construction, including design-led and contractor-led approaches, method planning, buildability assessment techniques, and case studies of best practices.
4. To analyse the role of stakeholders, clients, designers, contractors, regulators, and others, in the successful implementation of buildability principles.
5. To highlight the health and safety implications of buildability decisions, demonstrating how design choices affect workplace risks and site safety outcomes.
6. To articulate educational and capacity-building frameworks required to embed buildability competence within construction curricula and professional development programmes.
7. To demonstrate how buildability contributes to sustainability across environmental, economic, and social dimensions, aligning with global frameworks such as the Sustainable Development Goals and net-zero carbon agendas.
8. To identify emerging trends and future directions, including the role of advanced technology, digitalisation, modularisation, circular economy principles, and resilience planning in shaping buildability practice.

1.5 Scope of the Book

This book focuses on buildability as both a conceptual framework and a practical approach for improving construction outcomes. The scope encompasses:

1.5.1 Geographic Scope

Although it draws on diverse case studies, this book maintains a broad global perspective, addressing buildability challenges and research needs across a wide range of contexts.

1.5.2 Thematic Scope

This book covers foundational theory, practical applications, stakeholder roles, educational aspects, and sustainability pathways. Topics such as constructability, maintainability, health and safety, procurement, and construction methodology are addressed insofar as they relate directly to buildability.

1.5.3 Methodological Scope

This book draws on a synthesis of published literature, industry case studies, theoretical models, and conceptual frameworks. While not reporting empirical data, it engages with empirical findings from prior studies to support insights and interpretations.

1.5.4 Exclusions

This book does not delve deeply into specialised areas such as structural design calculations, financial modelling, or advanced engineering simulations, except where relevant to buildability principles.

1.6 Conceptual Orientation of the Book

Buildability in this book is viewed as an integrative construct that:

- Bridges design and construction.
- Improves project performance and sustainability.
- Enhances collaboration among stakeholders.
- Promotes rationalisation and simplification.
- Supports lifecycle efficiency and maintainability.
- Advances safety and risk management.
- Facilitates equitable and inclusive infrastructure development.

This book embraces both traditional and modern interpretations of buildability. Classical perspectives emphasise reducing construction complexity, standardising components, and optimising construction methods. Contemporary perspectives expand buildability to address environmental impact, carbon reduction, circularity, technological integration, prefabrication, health and safety, digital simulation, and value for society.

This orientation reflects the evolving nature of construction industries worldwide, where success increasingly depends on holistic thinking, integrated project delivery, and strategic planning that begins at the earliest stages of design.

1.7 Structure and Organisation of the Book

This book is organised into four major parts comprising ten chapters. Each part addresses a thematic dimension of buildability, progressing from foundational concepts to practical applications, educational needs, and future sustainability pathways.

Part I – Foundations of Buildability in Construction

This section establishes the theoretical basis of the monograph.

Chapter 2 explores the historical development of buildability globally, from early notions of constructability to modern institutionalisation, including key milestones, pioneers, and policy evolutions. It includes a conceptual clarification between buildability and constructability.

Chapter 3 provides conceptual and theoretical perspectives on buildability. It explains definitions, principles, attributes across project phases, and key influencing factors. The chapter also discusses theoretical lenses such as rationalisation, standardisation, transaction cost economics, signalling, and systems theory.

Part II – Applications in Construction Practice

This section addresses buildability in action.

Chapter 4 discusses buildability in procurement, design, construction methodology, maintainability, and early contractor involvement. It outlines how design decisions, method selection, and stakeholder engagement shape construction outcomes.

Chapter 5 focuses on buildability assessment, highlighting its concept, need, benefits, information requirements, tools, drivers, and challenges. It presents case studies demonstrating best practices.

Chapter 6 examines the relationship between buildability and health and safety. It explains how design decisions influence site risks and outlines strategies for improving safety outcomes through buildability.

Chapter 7 analyses the roles of key stakeholders, clients, designers, contractors, and regulatory bodies in promoting or constraining buildability implementation.

Part III – Education and Capacity Building

Chapter 8 addresses the skills, competencies, curriculum design, pedagogical approaches, and industry partnerships necessary for enhancing buildability education. It outlines how universities and professional bodies can strengthen buildability capabilities in the workforce.

Part IV – Sustainability and Future Pathways

Chapter 9 situates buildability within sustainable construction agendas. It highlights how buildability supports sustainability through efficient material use, reduced environmental impact, improved resilience, and alignment with net-zero and circular economy goals. The chapter also explores advanced technologies, modular construction, offsite manufacturing, BIM, digital twins, automation, and robotics.

Conclusion – Chapter 10

This concluding chapter synthesises insights and presents recommendations for policymakers, practitioners, educators, and researchers. It outlines future directions for buildability, including new research areas and strategic priorities for the next generation of construction practices.

1.8 Contribution of the Book

This book contributes to research and practice in several ways:

1.8.1 Theoretical Advancement

By synthesising classical and emerging perspectives, this book offers a comprehensive conceptual foundation for buildability. It integrates multiple theoretical lenses, enriching the understanding of how buildability influences project performance, stakeholder behaviour, and organisational outcomes.

1.8.2 Practical Insight

Through applied chapters and real-world examples, the monograph provides practitioners with actionable strategies for improving buildability in design, procurement, construction methodology, and safety management.

1.8.3 Educational Framework

It contributes to construction education by outlining competencies, pedagogical approaches, and curriculum structures essential for developing buildability-aware graduates.

1.8.4 Sustainability Lens

This book repositions buildability as a sustainability-enabling tool, demonstrating its critical role in resource efficiency, carbon reduction, resilience, and equitable development.

1.8.5 Policy Relevance

This book provides insights for regulatory bodies and policymakers seeking to improve national construction performance through buildability-led reforms.

1.9 Justification for Buildability as a Pathway to a Sustainable and Equitable Built Environment

A sustainable and equitable built environment requires more than structural adequacy or technical efficiency. It demands thoughtful design, transparent decision-making, lifecycle consciousness, and stakeholder inclusiveness. Buildability contributes to these goals by:

- Reducing waste and resource consumption and supporting circular economy principles.
- Lowering construction costs and timelines and enhancing affordability and equitable access to infrastructure.
- Improving quality and durability and ensuring long-term societal value.
- Enhancing safety and reducing risks to workers and surrounding communities.
- Supporting sustainable procurement and encouraging early contractor involvement and collaborative approaches.
- Facilitating adaptable and resilient structures, essential under climate change and urban pressures.

In contexts where access to infrastructure is uneven, such as in parts of Africa, Asia, and Latin America, buildability can help bridge inequity gaps by enabling cost-effective, durable, and inclusive design solutions. Integrating buildability into early design phases ensures that projects remain functional, maintainable, and environmentally responsible throughout their lifecycle.

1.10 Conclusion

This introductory chapter has presented the context, motivation, objectives, scope, and conceptual structure of this book. It has situated buildability as a multidimensional framework essential to delivering efficient, safe, sustainable, and equitable construction outcomes. The chapter has also outlined how this book is organised into thematic parts addressing foundational theories, practical applications, educational approaches, and sustainability futures.

The subsequent chapters build on this foundation, beginning with an exploration of the historical evolution of buildability in the global construction industry.

Bibliography

Djabarouti, J., & O'Flaherty, C. (2020). Architect and craftsperson: Project perceptions, relationships and craft. *Archnet-IJAR: International Journal of Architectural Research*, 14(3), 423–438. https://doi.org/10.1108/ARCH-01-2020-0010

Dosumu, O., & Aigbavboa, C. (2018). *Sustainable design and construction in Africa: A system dynamics approach* (1st ed.). Routledge. eBook ISBN 9781351212205. https://doi.org/10.1201/9781351212205

Dulaimi, M.F., Florence, Y.Y., & Ling, G.O. (2004). Engines for change in Singapore's construction industry: An industry view of Singapore's Construction 21 report. *Building and Environment*, 39(6), 699–711. https://doi.org/10.1016/j.buildenv.2004.01.011

Firoozi, A.A., Firoozi, A.A., Oyejobi, D.O., Avudaiappan, S., & Flores, E.S. (2024). Emerging trends in sustainable building materials: Technological innovations, enhanced performance, and future directions. *Results in Engineering*, 24, 103521. https://doi.org/10.1016/j.rineng.2024.103521

Grabowski, Z.J., Matsler, A.M., Thiel, C.L., McPhillips, L.E., Hum, R., Bradshaw, A., Miller, T., & Redman, C. (2017). Infrastructures as socio-eco-technical systems: Five considerations for interdisciplinary dialogue. *Journal of Infrastructure Systems*, 23(4), https://doi.org/10.1061/(ASCE)IS.1943-555X.0000383

Hassan, M.M., Tedong, P.A., Azlina, Khir, M., Shari, Z., Ponrahono, Z., Muhammad, & Sharifudin, P. (2023). Exploring the effects of the built environment on urban community wellbeing. *International Journal of Academic Research in Business and Social Sciences*, 13(10), 184–203. https://doi.org/10.6007/IJARBSS/v13-i10/18699

Mohd Nawi, M.N., Baluch, N., & Bahaudin, A.Y. (2014). Impact of fragmentation issue in construction industry: An overview. In *Conference: Building Surveying, Facilities Management and Engineering Conference (BSFMEC 2014). MATEC Web of Conferences*, 15(2), https://doi.org/10.1051/matecconf/20141501009

Osuizugbo, I.C. (2020). The need for and benefit of buildability analysis: Nigeria as a case study. *Journal of Engineering, Design and Technology*, 19(5), 1207–1230. https://doi.org/10.1108/JEDT-08-2020-0338

Osuizugbo, I.C. (2025). Buildability-led procurement: A new model for future construction projects delivery. In *Construction Innovation: Information, Process, Management.* https://doi.org/10.1108/CI-10-2023-0272

Tsao, K.E., Abdul-Samad, Z., Aziz, N.M., & Nagaratnam, B. (2021). Negotiation for managing conflict in construction industry with multicultural context: A systematic review of literature. *Journal of Design and Built Environment*, 21(3), 95–115. https://doi.org/10.22452/jdbe.vol21no3.7

Waqar, A., Houda, M., Khan, A.M., Qureshi, A.H., & Elmazi, G. (2024). Sustainable leadership practices in construction: Building a resilient society. *Environmental Challenges*, 14, 100841. https://doi.org/10.1016/j.envc.2024.100841

Part I

Foundations of Buildability in Construction

2 Historical Evolution of Buildability in the Global Construction Industry

2.1 Overview of Buildability as a Concept

Buildability, at its core, refers to the extent to which the design of a building or infrastructure project facilitates ease of construction, efficient utilisation of resources, and overall optimisation of project delivery outcomes. It describes the inherent capacity of a design to be translated into physical form with minimal difficulty, waste, ambiguity, or disruption. In this regard, buildability is not simply a by-product of good design; it is the result of intentional, informed, and collaborative decision-making that anticipates construction challenges and embeds construction knowledge at the earliest stages of project conception. As a concept, buildability sits at the intersection of design methodology, construction management, engineering innovation, and practical on-site operations. Its influence spans the entire project lifecycle, from conceptual planning through detailed design, procurement, and physical execution.

Although contemporary discourse often defines buildability in terms of reducing complexity, minimising waste, or improving constructability during the design phase, its underlying purpose is far deeper and more strategic. Buildability seeks to ensure that the architect's design intent aligns seamlessly with construction realities, resource constraints, environmental conditions, and the practical capabilities of the workforce. In doing so, it closes the traditional divide between design and construction, a divide that historically led to inefficiencies, cost overruns, rework, and miscommunication. The concept thus supports a fully integrated design-construction continuum, where design decisions are evaluated not only for aesthetic or functional merit but also for their implications on labour productivity, project scheduling, cost predictability, site logistics, safety performance, and long-term durability.

The importance of buildability becomes especially clear when one considers the traditional fragmentation of the construction industry. Historically, building production was marked by distinct professional silos: architects developed concepts and drawings; engineers applied structural calculations; contractors interpreted these documents and organised on-site activities; and trade specialists executed the physical work. Each group worked largely independently, and the flow of information between design and construction was often limited, sequential, and reactive

DOI: 10.1201/9781003747284-3

rather than integrated and proactive. This lack of coordination meant that construction challenges, such as difficult interfaces, impractical installation sequences, ambiguous detailing, or material incompatibilities, were often discovered only after work had begun on site. Such issues frequently resulted in delays, redesigns, budget escalations, and strained relationships among project stakeholders.

The rise of buildability as a formal concept emerged partly in response to these systemic inefficiencies. As construction projects became more complex, urban densities increased, and client expectations grew, the need for design solutions that were not only innovative but also construction-friendly became increasingly evident. Buildability thus evolved from an implicit design philosophy into a critical determinant of project success. Project owners and regulators began to recognise that early-stage design decisions have long-lasting and measurable impacts on outcomes such as construction cost, schedule adherence, labour efficiency, environmental performance, and overall project quality.

Over time, national and international research initiatives further reinforced the significance of buildability. Studies from the 1970s onward demonstrated that up to 70% of construction cost and productivity outcomes are influenced by design-stage decisions (e.g. CIRIA, 1983; Construction Industry Institute, 1986). Poor buildability was shown to correlate with high rates of rework, excessive manpower dependency, complicated sequencing, and avoidable material waste, issues that undermine a project's commercial viability and performance. In contrast, projects that incorporated buildability principles early typically experienced smoother workflows, more predictable costs, fewer disputes, and improved safety outcomes.

As a result, the industry witnessed the gradual institutionalisation of buildability. Countries such as the United Kingdom, Australia, Singapore, and the United States developed formal frameworks, guidelines, and rating systems to promote the integration of construction knowledge into design. Singapore's Buildability Framework, launched in 2001 and continuously refined, remains one of the most advanced examples, requiring mandatory buildability scores for building submissions to ensure that designs meet measurable standards of constructability, simplicity, and labour efficiency. Similar efforts in the UK, through CIRIA's buildability guidance, and in North America, through the Construction Industry Institute's constructability programmes, reflect a global recognition of the concept's strategic value.

Despite its increasing maturity, buildability is still often misunderstood as a narrow technical issue or an isolated design optimisation strategy. In reality, it is a multifaceted, cross-disciplinary principle that encompasses material selection, structural configuration, prefabrication opportunities, standardisation, sequencing logic, access and safety planning, temporary works integration, coordination of mechanical and electrical installations, and the overall clarity of design documentation. It also extends to workforce considerations: designs that reduce the need for excessive manual labour, specialised skills, or complex assembly can significantly improve productivity and mitigate risks arising from labour shortages.

Contemporary interpretations of buildability increasingly emphasise sustainability and lifecycle thinking. As global construction shifts toward greener materials, circular-economy principles, and low-carbon development, buildability now includes the design of structures that reduce waste, enable high levels of off-site fabrication, utilise renewable or recyclable materials, and support efficient disassembly or adaptive reuse. The alignment of buildability with sustainability reflects an evolution of the concept from a focus on construction efficiency alone to a broader concern with environmental performance and long-term value creation.

At the technological frontier, digitalisation has transformed the practice of buildability. Tools such as Building Information Modelling, digital twins, parametric design, and automated clash detection allow teams to evaluate buildability with unprecedented precision. Complex construction sequences can be simulated, potential clashes identified, and alternative design options tested long before physical work begins. Digital design management has therefore operationalised buildability in ways that were not possible in earlier decades, making it a quantifiable, iterative, and collaborative process rather than a subjective judgement.

Moreover, modern construction methodologies, such as Design for Manufacture and Assembly, modular construction, lean construction, 4D modelling, and integrated project delivery (IPD), have strengthened the strategic importance of buildability. These methodologies require early input from contractors, fabricators, and specialists; depend on simplification and standardisation of components; and emphasise the importance of efficient sequencing and reduced on-site assembly. As such, buildability has become a foundational design requirement for many large-scale, high-performance, and technologically advanced projects.

Despite these advancements, it is important to recognise that the concept of buildability is neither new nor limited to high-tech environments. Its roots stretch back centuries, embedded in the empirical knowledge of master builders, craftspeople, and engineers who understood that good design must accommodate the realities of available materials, construction techniques, climatic conditions, and labour skill levels. Ancient and medieval structures, from Roman aqueducts to Gothic cathedrals, demonstrate a profound awareness of buildability principles, even if the term itself did not exist. Throughout history, successful builders aligned architectural ambition with pragmatic construction strategies, balancing aesthetic goals with structural feasibility and resource efficiency.

Thus, although the vocabulary of buildability is modern, the underlying principles reflect long-standing traditions of practical knowledge, iterative learning, craftsmanship, and experience-based problem-solving. What distinguishes contemporary buildability from its historical manifestations is the degree of formalisation, scientific evaluation, and integration into regulatory frameworks. Modern buildability is evidence-based, systematically assessed, and supported by measurement systems that evaluate the ease with which a design can be constructed.

In this context, buildability can be understood as both a conceptual framework and a practical discipline:

- As a conceptual framework, it encourages designers, engineers, and contractors to adopt a holistic view of project delivery, recognising that decisions made in the design office have far-reaching consequences on the construction site.

- As a practical discipline, it guides specific design strategies, such as minimising structural complexity, standardising components, rationalising dimensions, planning logical sequences, improving access, and ensuring coordination between trades.

Underlying both perspectives is the belief that design and construction are not sequential but interdependent activities. Achieving high buildability requires early collaboration, open communication, and shared responsibility among all project participants. It also demands that construction knowledge, traditionally held by contractors and craftsmen, be integrated into the design process from the onset, enabling potential problems to be identified before they manifest on site.

In summary, buildability has evolved in response to changing industry needs, technological advancements, and global pressures for improved productivity, sustainability, and efficiency. It has grown from a tacit and intuitive design concern into a structured, measurable, and widely recognised framework that shapes how modern construction projects are conceived and executed. While its definition continues to adapt as new technologies and methodologies emerge, the core principle remains unchanged: aligning design intent with construction reality to build better, faster, safer, and more sustainably.

2.2 Origins and Early Conceptualisation of Buildability

The origins of buildability can be traced to the early industrial developments of the twentieth century, a period that marked the transition from traditional craft-based construction practices to more organised, large-scale, and technologically intensive building methods. As cities expanded rapidly and nations invested heavily in infrastructure, industrial plants, housing, and public institutions, the construction industry was confronted with unprecedented demands for speed, efficiency, and predictability. These pressures exposed the limitations of conventional design–construction processes, which were largely sequential, fragmented, and dependent on the tacit knowledge of craftsmen rather than systematic design principles. Buildability emerged in this context as an implicit but increasingly recognised need: the need for designs that were easier, faster, and more logical to construct.

2.2.1 Industrialisation and Method Standardisation

The early conceptualisation of buildability was profoundly influenced by industrialisation and the associated rise of scientific management and production engineering. Between the 1920s and 1950s, mass-production principles transformed global manufacturing sectors, particularly in automotive, shipbuilding, and steel production. Ideas such as standardisation, modularisation, workflow optimisation, and lean resource use began gaining traction as methods to improve productivity.

Although construction lagged behind manufacturing in adopting these practices, early innovators recognised that buildings too could benefit from standardised components and repeatable processes. The rise of modern prefabrication during this period, especially in Europe after World War II, illustrated the potential of reducing on-site complexity by designing components that could be manufactured and assembled efficiently. Structural elements with uniform dimensions, rationalised grid systems, interchangeable parts, and simplified detailing were early examples of what would later be formalised under buildability principles.

Industrial engineering philosophies, such as those championed by Henry Gantt, Frank and Lillian Gilbreth, and later W. Edwards Deming, also influenced construction. Their emphasis on workflow sequencing, waste reduction, motion efficiency, and quality control hinted at the broader need to rethink how construction projects were planned and executed. Even though they did not explicitly use the term "buildability," their contributions laid the analytical foundation for understanding how design decisions impact constructability.

2.2.2 Growing Separation between Designers and Constructors

Parallel to industrial advances, the twentieth century also witnessed increasing professionalisation within the built environment disciplines. Architecture, engineering, and contracting became distinct domains, each with its own educational pathways, regulatory structures, and practice norms. While specialisation fostered expertise, it also produced silos: designers became responsible for creating drawings and specifications, while contractors assumed responsibility for translating these into physical form.

This separation created persistent coordination challenges. Designers, often removed from construction realities, produced drawings that were difficult or inefficient to implement. Contractors, expected to interpret complex design information, frequently encountered ambiguous details, impractical sequencing assumptions, and insufficient allowances for access, logistics, or temporary works. As a consequence, many construction projects suffered from avoidable rework, delays, resource wastage, and escalating costs.

Early critiques of this fragmentation provide some of the first formal recognitions of the need for buildability. Notably, Sir Harold Emmerson's 1962 report, "Survey of Problems before the Construction Industries," was among the earliest systematic analyses to argue that the divide between design and construction was a root cause of inefficiencies in the industry. Although Emmerson did not use the term "buildability," he identified the same challenges the concept seeks to address: complicated communication pathways, lack of early contractor involvement (ECI), poor integration of construction expertise into design stages, and adversarial procurement processes. Emmerson recommended a new form of association between designers and builders, an early precursor to what is now known as constructability or design-construction integration.

His work stimulated subsequent debate within the UK construction sector, particularly highlighting that inefficiencies on site were frequently embedded in design decisions. This recognition became a crucial stepping stone toward the development of explicit buildability frameworks in the decades that followed.

2.2.3 1960s–1970s: Increasing Focus on Productivity Declines and Industry Fragmentation

By the 1960s and 1970s, concerns about construction inefficiency reached new heights. Many industrialised nations, including the UK, the United States, Japan, and Australia, began reporting stagnant or declining labour productivity in construction while productivity in manufacturing continued to rise sharply. The contrast highlighted fundamental differences between the two sectors: manufacturing embraced automation, standardisation, and continuous improvement, while construction remained heavily reliant on manual labour, bespoke designs, and site-based problem-solving.

Research groups, professional institutions, and government agencies began investigating these productivity gaps. Several influential studies concluded that poor buildability, though not yet labelled as such, was a key contributor. Design-related issues such as unnecessary complexity, inconsistent dimensions, lack of spatial coordination, impractical construction tolerances, and insufficient planning for component assembly were repeatedly identified as barriers to efficient construction.

During this period, scholars and practitioners increasingly called for a more systematic approach to evaluating the ease with which designs could be built. This early academic inquiry centred primarily on identifying design variables that influenced construction performance. Among the most frequently cited factors were:

- Standardisation of structural elements to simplify procurement, reduce formwork complexity, and facilitate repetitive construction actions.
- Rationalisation of dimensions, ensuring consistent modular grids, regular bay sizes, and uniform storey heights.
- Simplicity of detailing, including clear junctions, straightforward connections, and minimised bespoke components.
- Provision of access for assembly, maintenance, and inspection, recognising that a design must accommodate not just construction but the entire lifecycle of the building.
- Integration of temporary works considerations, acknowledging that practical construction support systems, such as scaffolding, shoring, and lifting zones, must be accounted for in design rather than improvised later.

These insights represented an important shift in thinking. Rather than treating construction as a separate stage, researchers began emphasising the upstream responsibility of design in shaping downstream outcomes. This period thus marks the intellectual foundation of buildability as a concept: a recognition that construction efficiency is largely determined long before a project reaches the site.

2.2.4 Emergence of Early Buildability Thinking across Global Regions

Several regions contributed to the early conceptualisation of buildability, each responding to their unique industrial contexts:

United Kingdom: The UK played a foundational role through the Emmerson Report (1962) and later the Banwell Report (1964), both of which criticised fragmented procurement processes and called for better design–construction integration. These critiques later influenced the work of CIRIA (Construction Industry Research and Information Association), which in the early 1980s developed some of the first structured buildability guidelines.

United States: In the U.S., parallel discussions around constructability began emerging by the late 1970s. Research initiatives at the University of Texas eventually led to the establishment of the Construction Industry Institute (CII) in 1983, which developed formal definitions, best practices, and industry-wide guidelines for integrating construction expertise into design.

Scandinavia and Japan: Innovations in prefabrication and modular systems, particularly in post-war Scandinavia and industrialised Japan, strengthened the recognition that design must be tightly integrated with manufacturing and assembly constraints. Such approaches implicitly underscored the need for buildable designs long before the term gained international popularity.

Australia: Australian researchers and government bodies began investigating design-for-construction principles in the 1970s, driven by labour shortages in major cities and the rapid expansion of public infrastructure. These early studies contributed to later national guidelines on buildability.

2.2.5 Buildability as an Under-Theorised but Emerging Concept

Despite growing recognition of its importance, buildability remained under-theorised during the mid-twentieth century. The term itself was not widely used, and where it did appear, it lacked a consistent definition. The concept functioned more as a practical intuition: designers and builders understood that simpler, standardised, and well-coordinated designs were easier to construct, but there were few formal tools, metrics, or frameworks for evaluating these qualities.

This absence of conceptual clarity reflected broader structural issues within the industry. Fragmented procurement models discouraged collaboration; contractual separation prevented contractors from influencing early design; and educational pathways for designers rarely included construction methodology or labour efficiency considerations. As a result, buildability principles often emerged informally through the experience of seasoned practitioners rather than through documented theory.

Nonetheless, the period between the 1960s and 1970s provided a fertile ground for the concept to mature. Empirical evidence linking design decisions to construction outcomes became more widely circulated. Case studies demonstrated the cost savings associated with standardised grids, simple reinforcement layouts, and rationalised detailing. Technological advancements, such as early computer-aided design systems and improved structural analysis tools, provided new ways for designers to examine construction implications more systematically.

By the late 1970s, many national and international bodies were openly calling for a transformation of design practices. The idea that designs should explicitly consider construction efficiency became a recurring theme in industry reports, academic conferences, and policy discussions. Even though formal buildability frameworks were still emerging, the conceptual seeds had been firmly planted.

2.2.6 Conclusion: Foundations for Systematisation

The early origins and conceptualisation of buildability were shaped by intertwined historical forces: industrialisation, professional fragmentation, rising project complexity, and concerns about declining productivity. These factors collectively highlighted the inadequacies of traditional design–construction separation and underscored the need for more integrated and construction-informed design processes.

Although the term "buildability" had not yet achieved widespread usage in the mid-twentieth century, its underlying principles, standardisation, simplification, coordination, and alignment of design intent with construction reality, were already taking shape. Early critiques by influential figures such as Sir Harold Emmerson played a pivotal role in articulating the systemic challenges facing the industry and advocating for greater collaboration between designers and constructors.

This period laid the intellectual groundwork for the formal frameworks, guidelines, and rating systems that emerged in the 1980s and beyond. Buildability transitioned from an under-theorised intuition to a central organising principle of modern construction design, setting the stage for the more structured institutionalisation explored in subsequent sections.

2.3 Expansion and Institutionalisation of Buildability

The 1980s and 1990s marked a transformative period for the global construction industry, characterised by increasing scientific inquiry, structural reforms, and technological advancements. Within this broader landscape, the concept of buildability, long understood informally by practitioners, began to crystallise into a formalised body of knowledge. This institutionalisation was driven by a global shift toward performance benchmarking, construction innovation, IPD, and productivity enhancement. As a result, buildability evolved from an intuitive design philosophy into a structured, evidence-based discipline supported by frameworks, guidelines, regulatory tools, and industry-led research initiatives.

This period of expansion can be understood through three interrelated dynamics: (1) the emergence of formal frameworks and guidelines, (2) global drivers sustaining widespread adoption, and (3) the integration of buildability principles into contemporary construction practices.

2.3.1 Emergence of Formal Frameworks and Guidelines

Prior to the 1980s, buildability existed mostly as a practice-informed intuition. While the importance of integrating construction knowledge into design had been acknowledged, particularly following critiques from the 1960s such as the Emmerson (1962) and Banwell (1964) reports in the UK, no consistent methodology or industry-wide standard for evaluating buildability existed. The 1980s marked the beginning of formal institutionalisation as national and international bodies sought to transform buildability into a measurable performance attribute. This created the foundation for global adoption and standardised integration of buildability into design and procurement processes.

Construction Industry Institute (CII), United States: One of the most significant milestones in formalising buildability occurred with the establishment of the CII at the University of Texas at Austin in 1983. CII introduced the term *constructability*, closely related to buildability, and defined it as the optimal use of construction knowledge and experience in planning, design, procurement, and field operations to achieve overall project objectives. CII's contributions included: (1) A seminal publication providing structured guidelines on integrating construction knowledge into design. (2) A practical methodology outlining when and how constructability reviews should occur. (3) Encouraging early contractor involvement (ECI) through structured workshops and design checkpoints. CII's research formalised buildability/constructability into a systematic set of principles and demonstrated, through case studies, its measurable impact on cost, time, and quality performance. Their work became internationally influential, providing a replicable model for institutionalising buildability.

Building Research Establishments and CIRIA, United Kingdom: The UK followed a parallel path to formalisation. Building on earlier critiques of industry fragmentation, organisations such as the Building Research Establishment (BRE) and the Construction Industry Research and Information Association (CIRIA) began systematically studying buildability in the late 1970s and early 1980s. Key UK contributions included: (1) Development of early buildability guidance documents, emphasising design factors that influence ease of construction. (2) Integration of buildability considerations into design performance evaluations and industry best-practice manuals. (3) Establishment of interdisciplinary working groups exploring how procurement systems and contractual arrangements affect buildability. CIRIA's publications, such as *Buildability: An Assessment* (1983) and later design integration guidelines, were among the first to formally define buildability and offer detailed criteria for measuring the constructability of building designs. These efforts significantly shaped global academic research on the subject.

Singapore's CIDA/BCA and the Buildable Design Appraisal System (BDAS): One of the most influential examples of national-level institutionalisation came from Singapore, where labour shortages, land constraints, and the need for rapid urban development prompted a strong focus on improving construction productivity. Singapore's Construction Industry Development Board (CIDB), which later became the Building and Construction Authority (BCA), introduced the BDAS in 2001. BDAS became one of the world's first mandatory regulatory frameworks assessing buildability at the design approval stage. Key features of BDAS include: (1) A scoring system evaluating the buildability of architectural, structural, and MEP designs. (2) Incentives and regulatory requirements for higher buildability scores.

Integration of labour-efficiency models into the approval process. (3) Promotion of standardised components, rationalised layouts, and simplified detailing. BDAS institutionalised buildability by embedding it directly in the statutory review process, making constructability a central requirement rather than an optional design consideration. Singapore's success demonstrated how policy instruments can drive systemic improvements in design and construction efficiency.

Other International Developments: Throughout the 1980s and 1990s, several additional regions contributed to the institutionalisation of buildability. Australia introduced early buildability assessment guidelines through its industry development bodies and academic institutions. Scandinavian countries, especially Finland and Sweden, formalised buildability principles within prefabrication-led housing and industrial construction programmes. Japan's construction giants (e.g., Shimizu, Kajima, Obayashi) integrated constructability into their design–build and off-site manufacturing systems, supported by advanced automated construction technologies. Collectively, these developments accelerated a shift from buildability as informal knowledge toward buildability as a codified, organisation-wide design management practice.

2.3.2 Global Drivers of Adoption

The institutionalisation of buildability throughout the late twentieth century did not emerge in a vacuum. Rather, it was shaped by powerful global drivers that exposed the limitations of traditional design–construction separation and highlighted buildability as a strategic solution for emerging industry-wide challenges.

Rising Construction Costs and Productivity Pressures: From the 1970s onward, most industrialised nations experienced declining labour productivity in construction, even as productivity in manufacturing continued to rise. Construction projects became increasingly expensive, with large proportions of cost overruns attributable to design deficiencies, uncoordinated details, and inefficient construction workflows. Buildability offered a means to counter these trends by: (1) reducing rework, (2) simplifying construction processes, (3) improving sequencing and logistics, (4) minimising labour-intensive operations, and (5) enabling more predictable cost performance. The need for productivity improvement became a key motivator for formal buildability frameworks.

Increased Project Complexity: The late twentieth century saw unprecedented growth in: (1) high-rise building construction, (2) long-span structures, (3) underground transit systems, (4) advanced mechanical and electrical services, and (5) large-scale infrastructure megaprojects. Such complexity meant that construction interfaces multiplied, tolerances became tighter, and the consequences of design deficiencies intensified. Buildability frameworks helped manage this complexity by encouraging: (1) better coordination between disciplines, (2) rationalisation of layouts, (3) standardisation of details across trades, and (4) prioritisation of constructible solutions that reduced risk and uncertainty.

Demands for Shorter Construction Timeline: Clients and governments increasingly demanded faster project delivery, particularly in rapidly urbanising regions. The rise of global competition also placed pressure on developers and contractors to deliver more in less time. Buildability facilitated: (1) acceleration of construction schedules, (2) reduction in site disruptions, (3) smoother critical-path activities, and (4) improved prefabrication and assembly strategies.

Advancements in Construction Technology: Advancements in construction technology have made buildability both more essential and more achievable. Prefabrication and industrialised building systems demand standardised design layouts and modular coordination, which are fundamental principles of buildability. Automation and robotics further reinforce this need, as these systems operate most effectively with simplified, repetitive geometry and consistent detailing. Likewise, early computer-aided design (CAD) improved designers' ability to analyse constructability during the design phase, laying the groundwork for later innovations such as BIM.

Growing Emphasis on Sustainability and Lifecycle Thinking: By the 1990s, sustainable development frameworks emphasised the need to minimise waste, reduce embodied energy, optimise material utilisation, prolong building lifespan, and improve maintainability. Buildability supported these goals by promoting simple and efficient material use, integrating lifecycle maintenance considerations into the design process, enabling deconstruction or adaptive reuse through construction-friendly layouts, and facilitating low-waste off-site manufacturing strategies. This strong alignment between sustainability and buildability further reinforced the case for its institutionalisation.

2.3.3 Integration into Contemporary Practices

Entering the twenty-first century, buildability had fully transitioned from an emerging concept to a core component of modern construction practice. Its principles are now deeply embedded in contemporary project delivery models, digital tools, and industrialised building systems.

BIM and Digital Design Integration: BIM revolutionised buildability assessment by allowing construction logic to be simulated virtually. BIM-enabled workflows make it possible to evaluate: accessibility for installation, trade coordination, clash detection, crane and logistics planning, temporary works integration, and sequencing simulation through 4D modelling. This shift elevated buildability from a checklist-based method to a dynamic, model-driven process. Digital twins further enhanced this capability by enabling real-time constructability analysis based on actual site conditions.

Design for Manufacture and Assembly (DfMA): DfMA represents one of the strongest modern manifestations of buildability. It focuses on off-site production, standardised components, repeatable manufacturing processes, and minimised on-site assembly. Buildability principles form the backbone of DfMA, ensuring that designs are compatible with factory environments and allow rapid, accurate assembly on site. This approach is now widely used in: modular housing, precast concrete systems, volumetric construction, hospital and school programmes, and infrastructure components (e.g., bridges and stations).

Lean Construction: Lean construction, derived from Toyota's lean production philosophy, reinforces buildability's emphasis on simplification, waste reduction, continuous flow, and value maximisation. Lean tools such as Last Planner®

System, value stream mapping, and Just-In-Time delivery complement buildability by ensuring that design and construction processes align to minimise inefficiency.

Integrated Project Delivery (IPD): IPD models institutionalise buildability by embedding contractors, fabricators, and specialists into early design phases. This ensures that: constructability knowledge informs design decisions, risks are shared among parties, and collaboration replaces adversarial procurement systems. IPD is now widely used in complex sectors such as healthcare, aviation, and large public infrastructure.

Continuous Performance Measurement: Modern construction increasingly relies on design quality indicators (DQIs), constructability rating systems, digital audit tools, and productivity measurement frameworks. These instruments incorporate buildability criteria, making the concept both measurable and accountable.

2.3.4 Conclusion: Buildability as a Modern Expectation

By the end of the twentieth century and into the twenty-first, buildability had evolved from an informal practitioner insight into a formalised, globally recognised discipline. The establishment of guidelines by bodies such as CII, CIRIA, BRE, and BCA provided structured methodologies, while global pressures, rising costs, complexity, technological change, and sustainability imperatives accelerated adoption.

Today, buildability is not optional but an expected attribute of high-performance construction. It is embedded in digital design processes, industrialised production systems, lean methodologies, procurement strategies, and regulatory frameworks. As a result, buildability serves as a foundational principle in achieving efficient, predictable, safe, and sustainable construction outcomes in the modern era.

2.4 Conceptual Clarification between Buildability and Constructability

The concepts of buildability and constructability appear frequently in construction management literature, often interchangeably, leading to conceptual ambiguity. While both terms express a shared intention, enhancing the efficiency of delivering built assets, their historical, regional, and methodological foundations exhibit important distinctions. Buildability has its intellectual roots primarily in the Commonwealth tradition, with early developments centred around the United Kingdom, Australia, and later Singapore. Constructability, by contrast, evolved more explicitly through the influence of U.S. construction management research and the Construction Industry Institute. A clear differentiation between these terms is therefore necessary, not in order to divide the concepts artificially, but to illuminate their complementary roles in shaping integrated, design-construction workflows. The following subsections elaborate on these distinctions, their overlaps, and their convergence in contemporary global construction practice.

2.4.1 Buildability

Buildability is traditionally conceptualised as a design-driven approach that emphasises the extent to which building design decisions facilitate safe, efficient, and economical construction. Several scholars describe buildability as fundamentally concerned with the quality and appropriateness of design information (Ferguson, 1989; Wong et al., 2007). In its foundational definition, CIRIA (1983) describes buildability as "the extent to which the design of a building facilitates ease of construction," underscoring its design-centric orientation. The emphasis is on ensuring that design intent does not create unnecessary complications during construction and that design decisions actively consider the practicalities of assembly, sequencing, and integration of building components.

Design Dominance and Early Adoption: Historically, buildability emerged through critiques of poor design-construction integration in the UK construction sector. Early studies, such as those by the Chartered Institute of Building (CIOB, 1982) and the Building Research Establishment (BRE), pointed to the negative consequences of overly complex design detailing, insufficient coordination among consultants, and limited contractor input during the conceptual phases of projects. As a result, buildability research centred on identifying design factors that had a disproportionate influence on construction difficulty, including: (1) dimensional coordination and rationalisation, (2) structural standardisation, (3) simplicity of detailing, (4) repetition and modularisation, (5) simplicity in joints and connections, (6) ease of temporary works integration, and (7) site access and material handling considerations. This orientation established buildability as a pre-construction discipline, largely governed by architects and engineers rather than contractors.

Regional Usage and Institutionalisation: Buildability is widely used across Commonwealth jurisdictions including the UK, Australia, New Zealand, and Singapore. The Singaporean case is especially significant, as the Building and Construction Authority (BCA) formalised buildability through the Buildable Design Appraisal System (BDAS) (BCA, 2005), which is used as a mandatory scoring mechanism for building approvals. BDAS allocates points based on choices of structural systems, wall systems, and other design strategies that improve construction efficiency, highlighting buildability's firmly design-embedded character. In the UK, buildability is incorporated into guidance documents such as CIRIA Report 166 and various BRE publications, focusing on encouraging designers to consider buildability as part of good professional practice.

Scope and Emphasis: The distinguishing characteristics of buildability include: (1) Emphasis on the design phase rather than the entire project life cycle. (2) Prioritisation of design simplification and reduction of on-site complexity. (3) Use of design tools and checklists aimed at assessing construction implications. (4) Association with regulatory frameworks in certain jurisdictions (e.g., Singapore). (5) Adoption within design professions such as architecture and engineering. While buildability does influence site operations, its primary function lies in shaping design decisions that pre-empt construction inefficiencies.

2.4.2 *Constructability*

Constructability is a broader, more comprehensive concept than buildability, originating largely in the United States through research sponsored by the Construction Industry Institute in the 1980s. CII (1986) defines constructability as "the optimum use of construction knowledge and experience in planning, design, procurement, and field operations to achieve overall project objectives." This definition reflects three major differences from buildability: (1) Constructability spans the entire project life cycle, from early planning to commissioning. (2) It emphasises proactive contractor involvement, not just design rationalisation. (3) It integrates construction expertise into decision-making across procurement, logistics, sequencing, site management, safety, and resource planning.

Life-Cycle Orientation: While buildability is concentrated at the design stage, constructability encompasses: project initiation and feasibility, conceptual and detailed design, procurement strategies, material logistics and supply-chain coordination, construction sequencing and phasing, risk assessment, temporary works planning, labour productivity, and site organisation commissioning and turnover. This life-cycle orientation stems from research linking ECI to improved project outcomes (CII, 1993; Glavinich, 1995). Projects adopting constructability principles typically demonstrate improvements in cost certainty, schedule reliability, safety performance, and labour productivity.

Industry Influence and Standardisation: Constructability became institutionalised in the U.S. through the formation of CII Constructability Task Forces, the publication of constructability manuals, and widespread adoption by industrial clients, particularly in petrochemical, energy, and heavy infrastructure sectors. CII's structured review model includes constructability reviews at multiple project stages, supported by formal documentation and checklists. Other countries, including Canada, Brazil, and parts of Europe, adopted constructability frameworks through practitioner guidelines and project management standards. The Australian construction industry also embraced constructability, often blending it with buildability, leading to hybrid usage of the terms.

Scope and Emphasis: Constructability's distinguishing elements include: (1) Life-cycle perspective (planning to completion) (2) Strong linkage to construction management practices (3) Focus on contractor expertise and early involvement (4) Integration with procurement, logistics, scheduling, and risk management (5) Development from a construction-driven rather than design-driven tradition. Compared to buildability, constructability is more comprehensive, strategic, and management-oriented. Table 2.1 provides a summary clarifying the conceptual distinctions between buildability and constructability. The continuum model that illustrates how buildability and constructability interact across project phases is shown in Figure 2.1.

Figure 2.1 illustrates how buildability and constructability interact across project phases: planning, design, procurement, and construction. Figure 2.1 contains

Table 2.1 Conceptual distinctions between buildability and constructability

Dimension	*Buildability*	*Constructability*
Primary focus	Design optimisation to facilitate ease of construction	Integration of construction expertise into all project phases
Phase emphasis	Mainly design phase (conceptual + detailed)	Entire project life cycle (planning → design → procurement → construction → commissioning)
Historical origin	UK/Commonwealth; BRE, CIRIA work in 1970s–1980s	United States; CII programmes from mid-1980s
Key drivers	Simplification, standardisation, modularity	Project integration, planning efficiency, logistics, and safety
Typical tools	Buildability checklists, BDAS scoring, design rationalisation frameworks	Constructability reviews, CII constructability principles, stage-gate review systems
Profession most involved	Architects and engineers	Contractors, construction managers, multidisciplinary teams
Regulatory presence	Strong in Singapore (BDAS); guidance in UK and Australia	Less regulatory; driven by industry best practices
Scope of influence	Design quality, simplicity, detailing, modularity	Sequencing, procurement, logistics, risk, temporary works, and site operations
Degree of contractor involvement	Limited or early-measured involvement	High – requires formal early contractor involvement (ECI)
Typical outcomes	Reduced design complexity and improved construction efficiency	Improved overall project performance, cost/schedule certainty
Conceptual nature	Design-centric	Management-centric and holistic
Overlap areas	Simplicity, waste reduction, safety, coordination	Simplicity, waste reduction, safety, and coordination

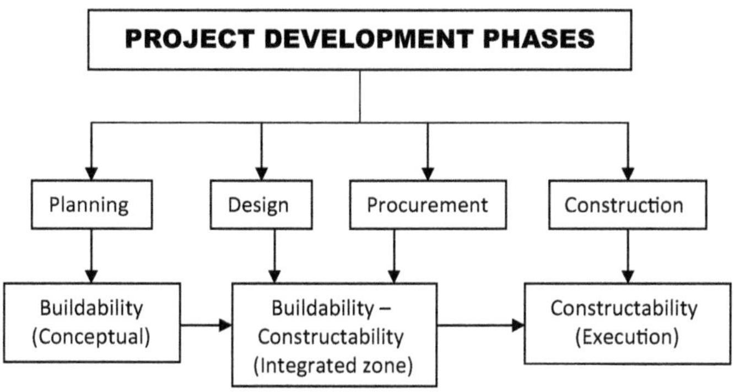

Figure 2.1 Buildability–constructability continuum model.

three defined zones, namely, (1) Buildability zone – represents early-stage design influences where decisions regarding simplicity, standardisation, and modularity are made. (2) Integration zone – depicts where both designer-led buildability considerations and contractor-driven constructability knowledge merge, particularly during late design and procurement. (3) Constructability zone – highlights full construction-phase optimisation, including sequencing, logistics, temporary works, and site operations.

The continuum model (Figure 2.1) clarifies that buildability and constructability are not competing concepts but mutually reinforcing, each dominating different parts of the project lifecycle. This visual reinforces the growing need for integrated workflows in modern digital and collaborative project delivery environments.

2.4.3 Overlaps and Complementarity

Despite their distinctions, buildability and constructability share a common foundation enhancing the alignment between design intent and construction execution. Both concepts emerged as responses to long-standing fragmentation in the construction industry, which hindered communication between design professionals and contractors. As industries have adopted more collaborative models, the lines between buildability and constructability have blurred.

Shared Principles: Both buildability and constructability emphasise (1) the integration of construction expertise into design decisions; (2) reducing unnecessary complexity in building systems; (3) optimising materials, methods, and sequencing; (4) enhancing productivity, quality, safety, and cost control; (5) eliminating waste and improving workflow continuity; and (6) improving coordination among multidisciplinary teams. Contemporary studies confirm that whether labelled buildability or constructability, the same underlying principles enhance project outcomes (Osuizugbo, 2025a, 2025b; Wimalaratne et al., 2021, 2023).

Convergence through Modern Technologies and Practices: Digital transformation has further blurred distinctions between the two concepts. Building Information Modelling provides platforms for visualising, simulating, and coordinating construction processes early in design, effectively merging buildability (design phase) with constructability (execution phase). Automated clash detection, 4D sequencing, and virtual construction modelling embody both concepts simultaneously. Design for Manufacture and Assembly requires designers to consider fabrication and assembly constraints early in the design process, enhancing buildability while incorporating constructability principles across logistics and supply chains. Lean construction eliminates waste and optimises workflow during both design and construction, representing a holistic approach that integrates both buildability and constructability. Integrated Project Delivery and Early Contractor Involvement procurement models place contractors alongside designers during early phases, strengthening the design-construction continuum. This convergence reflects the systemic nature of modern construction, where design decisions cannot be isolated from their execution implications.

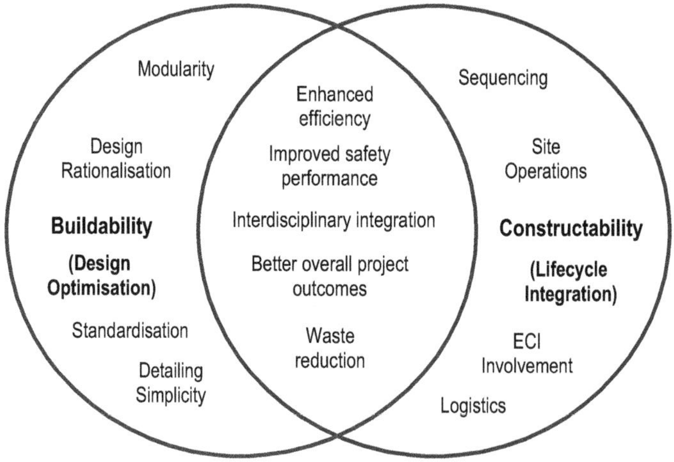

Figure 2.2 Venn diagram linking buildability and constructability.

Contemporary Global Practice: Today, many regions use the terms interchangeably because of the following: (1) Technological integration (BIM, DfMA, digital twins) creates unified workflows. (2) Collaborative procurement models make early contractor involvement common. (3) Sustainability requirements emphasise life-cycle thinking. (4) Prefabrication and modular systems demand simultaneous design and logistical planning. Despite this convergence, distinctions remain relevant for academic study, legal frameworks, and region-specific policy implementation. Buildability retains greater regulatory influence in Singapore and the UK, while constructability has deeper institutional frameworks in the U.S. and Canada.

Figure 2.2 provides a conceptual Venn diagram illustrating the areas where buildability and constructability overlap. The two circles represent the distinct emphases of each concept, one focused on design, the other on construction, while the intersection emphasises their shared goals, including: improved safety performance, enhanced efficiency, interdisciplinary integration, waste reduction, and better overall project outcomes. Figure 2.2 demonstrates that while each concept has unique origins and applications, they converge around common principles essential for successful project delivery. This visual representation helps communicate the unified nature of modern practice, where digital tools, ECI, and collaborative procurement models blur traditional boundaries between design and construction roles.

2.5 Chapter Conclusion

This chapter has traced the historical development of buildability from an implicit practical concern to a recognised and increasingly institutionalised concept within the global construction industry. By examining its origins, drivers, and evolving interpretations, this chapter demonstrates that buildability did not emerge in isolation but as a response to structural changes, productivity challenges, and growing fragmentation within construction systems worldwide.

The early conceptualisation of buildability can be linked to the broader forces of industrialisation and method standardisation. As construction processes became more complex and specialised, the separation between designers and constructors widened, often leading to inefficiencies and coordination failures. During the 1960s and 1970s, concerns over declining productivity, cost overruns, and adversarial contracting practices intensified the need for improved integration between design and construction. Across different global regions, early buildability thinking began to surface as practitioners and policymakers recognised that many project inefficiencies originated from design decisions that failed to account for construction realities. Although initially under-theorised, these early efforts laid the groundwork for more systematic approaches.

The expansion and institutionalisation phase marked a turning point in the formal recognition of buildability. Governments, industry bodies, and professional institutions began developing structured frameworks, guidelines, and assessment tools to embed constructability considerations into project delivery. Global drivers, including increasing project complexity, competitive pressures, regulatory reforms, technological advancement, and demands for improved safety and sustainability, have accelerated adoption. Buildability gradually transitioned from a corrective practice applied reactively to a proactive expectation integrated into procurement strategies, collaborative contracting models, and digital design environments. This evolution reflects a broader industry shift toward integration, lifecycle thinking, and performance-based project delivery.

The chapter also clarified the conceptual distinction between buildability and constructability. While often used interchangeably, buildability generally emphasises the extent to which design facilitates ease and efficiency of construction, whereas constructability focuses more broadly on the optimal integration of construction knowledge and experience throughout the project lifecycle. Despite subtle differences in emphasis, both concepts share a common objective: improving project outcomes through better coordination, simplified design, and early stakeholder involvement. Their overlaps and complementarities underscore the importance of holistic and integrated approaches rather than rigid terminological distinctions.

In conclusion, the historical evolution of buildability reflects the construction industry's ongoing effort to address fragmentation, inefficiency, and risk through greater integration and systematisation. From its early roots in industrial rationalisation to its current status as a modern industry expectation, buildability has evolved into a central pillar of effective project delivery. Understanding this historical trajectory provides critical insight into why buildability remains essential today and highlights the continuing need for innovation, collaboration, and institutional support to sustain its advancement in the global construction landscape.

Bibliography

BCA (Building and Construction Authority) (2005). *Code of practice on buildable design.* BCA.

Banwell, H. (1964). *The placing and management of contracts for building and civil engineering work: Report of the committee.* Her Majesty's Stationery Office.

Chartered Institute of Building (1982). *Code of practice for buildability.* Chartered Institute of Building.

Construction Industry Institute (CII) (1986). *Buildability: A primer.* University of Texas at Austin.

Construction Industry Institute (CII) (1993). *Preview of constructability implementation.* CII.

Construction Industry Research and Information Association (CIRIA) (1983). *Buildability: An assessment.* CIRIA Publications, Special Report No. 26.

Emmerson, H. (1962). *Survey of problems before the construction industries. A report prepared for the Ministry of Works.* London: HMSO.

Ferguson, I. (1989). *Buildability in practice.* Mitchell's Professional Library.

Fisher, D.J., & Rajan, N. (1996). Automated buildability: Analyze of work-zone traffic control planning. *Journal of Construction, Engineering and Management,* 122(1), 36–43.

Glavinich, T.E. (1995). Improving constructability during design phase. *Journal of Architectural Engineering,* 1(2). https://doi.org/10.1061/(ASCE)1076-0431(1995)1:2(73)

Griffith, A. (1987). An investigation into factors influencing buildability and levels of productivity for application to selecting alternative design solutions – A preliminary report. In *Proceedings of the 5th CIB W65 Symposium on Organisation and Management of Construction* (Vol. 2, pp. 646–657). Chartered Institute of Building.

Hyde, R. (1995). Buildability as a design concept for architects: A case study of laboratory buildings. *Engineering, Construction and Architectural Management,* 2(1), 45–56. https://doi.org/10.1108/eb021002

Illingworth, J.R. (1984). Buildability – Tomorrow's need. *Building Technology and Management,* 16–19.

Jarkas, A.M. (2010). Analysis and measurement of buildability factors affecting edge formwork labour productivity. *Journal of Engineering Science and Technology Review,* 3(1), 142–150.

Mydin, S.H., Zin, R.M., Abdi Majid, M.Z., Zahidi, M., & Memon, A.H. (2011). Buildability attributes at design phase in Malaysian building construction. *International Journal of Sustainable Construction Engineering & Technology,* 2(1), 24–43.

Osuizugbo, I.C. (2020a). Improving the performance of building construction firms through addressing the gap of building production management: A new production model approach. *Journal of Engineering, Project, and Production Management,* 10(1), 50–63.

Osuizugbo, I.C. (2020b). The need for and benefit of buildability analysis: Nigeria as a case study. *Journal of Engineering, Design and Technology,* 19(5), 1207–1230. https://doi.org/10.1108/JEDT-08-2020-0338

Osuizugbo, I.C. (2025a). Buildability-led procurement: A new model for future construction projects delivery. In *Construction Innovation: Information, Process, Management.* https://doi.org/10.1108/CI-10-2023-0272

Osuizugbo, I.C. (2025b). Measures for improving the buildability of building designs in construction industry. *Construction Innovation,* 25(2), 490–509. https://doi.org/10.1108/CI-08-2022-0223

Osuizugbo, I.C., Ezeokoli, F.O., Okolie, K.C., & Olojo, A.D. (2024). Towards a successful implementation of buildability assessment in construction industry: A conceptual framework. *Journal of Engineering, Design and Technology*. https://doi.org/10.1108/JEDT-08-2023-0359

Osuizugbo, I.C., & Ojelabi, R.A. (2020). Building production management practices in the construction industry in Nigeria. *Engineering Management in Production and Services*, 12(2), 56–73. https://doi.org/10.2478/emj-2020-0011

Osuizugbo, I.C., & Oshodi, O.S. (2023). Buildability attributes for improving the practice of construction management in Nigeria. *Built Environment Project and Asset Management*, 13(2), 270–289. https://doi.org/10.1108/BEPAM-03-2022-0051

Wimalaratne, P.L.I., Kulathunga, U., & Gajendran, T. (2021). Comparison between the terms constructability and buildability: A systematic literature review. In: Sandanayake, Y.G., Gunatilake, S., & Waidyasekara, K.G.A.S. (Eds), *Proceedings of the 9th World Construction Symposium*, 9–10 July 2021, Sri Lanka, pp. 196–207. https://doi.org/10.31705/WCS.2021.17

Wimalaratne, P.L.I., Kulatunga, U., & Gajendran, T. (2023). Buildability' in the digital age: A phenomenological discourse of industry practitioners' perceptions. *Buildings*, 13(11), 2870. https://doi.org/10.3390/buildings13112870

Wong, F.W.H., De Saram, D.D., Lam, P.T.I., & Chan, D.W.M. (2006a). A compendium of buildability issues from the viewpoints of construction practitioners. *Architectural Science Review*, 49(1), 81–90. https://doi.org/10.3763/asre.2006.4910

Wong, F.W.H., Lam, P.T.I., Chan, A.P.C., & Chan, E.H.W. (2006b). A review of buildability performance in Hong Kong and strategies for improvement. *Surveying and Built Environment*, 17(2), 37–48.

Wong, F.W.H., Lam, P.T.I., Chan, E.H.W., & Wong, F.K.W. (2006c). Factors affecting buildability of building designs. *Canadian Journal of Civil Engineering*, 33, 795–806, https://doi.org/10.1139/L06-022

Wong, F.W.H., Lam, P.T.I., Chan, E.H.W., & Shen, L.Y. (2007). A study of measures to improve constructability. *International Journal of Quality & Reliability Management*, 24(6), 586–601. https://doi.org/10.1108/02656710710757781

3 Buildability

Concept and Theoretical Perspectives

3.1 Definitions of Buildability

Buildability has long been recognised as a central concern in construction management, reflecting the need to harmonise design intentions with the practical realities of construction. Because the concept lies at the intersection of design, engineering, project management, and site operations, it has attracted considerable scholarly attention. Over the past four decades, numerous definitions have been proposed, each emphasising different aspects of the design–construction interface. As a result, the literature reflects a rich but sometimes fragmented understanding of what buildability entails and how it should be operationalised in practice.

In general terms, buildability refers to how well a building design supports efficient, safe, and cost-effective construction. While this broad interpretation is widely accepted, authors vary in the specific dimensions they prioritise. Early work by CIRIA (1983) placed buildability within the context of design rationalisation, arguing that the ease with which construction can be executed is heavily dependent on early design decisions. CIRIA's definition underscores the idea that buildability is not a construction-phase issue alone but one rooted in thoughtful design development and interdisciplinary coordination.

Similarly, Illingworth (1984) emphasised the role of design simplification, suggesting that a buildable design is one that minimises unnecessary complexity. This view links buildability to the reduction of waste, errors, and rework, issues that persistently challenge construction productivity. Griffith (1987) further expanded the concept by framing buildability as a mechanism to enhance communication between designers and constructors. His work recognised that the fragmentation of the construction industry often creates disconnects between those who design and those who build, making buildability a tool for bridging this gap.

Ferguson (1989) contributed to the discussion by highlighting the relationship between buildability and construction performance. He argued that a design with high buildability not only improves productivity but also reduces labour intensity and enhances safety on-site. This focus on labour efficiency resonates with broader industry concerns about skills shortages, fluctuating labour markets, and the importance of standardisation in modern construction.

DOI: 10.1201/9781003747284-4

Building on these foundations, Moore (1993) and Chen et al. (1993) offered more comprehensive definitions, incorporating both managerial and technical dimensions. They viewed buildability as the result of deliberate integration between design strategies and construction methods, enabled through early contractor involvement, effective planning, and design optimisation. Their work links buildability to decision-making processes, implying that it is not merely a characteristic of a design but an outcome of collaborative project management practices.

Fisher and Rajan (1996) took this further by conceptualising buildability within the broader framework of constructability, often treated as a closely related or interchangeable term. While constructability is more commonly used in North American literature, particularly through the influential work of the Construction Industry Institute (CII) in 1986, both terms share the fundamental notion of improving project outcomes through better integration of design and construction knowledge. CII (1986) defined constructability as the optimal use of construction knowledge and experience in planning, design, procurement, and field operations. Although some authors differentiate between the two terms, suggesting that buildability focuses more on design and constructability more on the entire project lifecycle, many regard them as synonymous or overlapping concepts.

Bamisile (2004) and Lam et al. (2006) provided more contemporary interpretations that reflect the increasing complexity of modern construction projects. They emphasised that buildability involves not only simplifying design but also considering materials selection, technological compatibility, site logistics, supply chain integration, and regulatory constraints. These authors argue that buildability should be viewed as a holistic concept that spans the entire project process rather than as an isolated design criterion.

Across these definitions, several key themes consistently emerge: (1) *Design Simplicity and Rationalisation*: A buildable design minimises unnecessary complexity, facilitates straightforward construction sequences, and reduces opportunities for errors. (2) *Integration of Construction Knowledge into Design*: Early contractor involvement and multidisciplinary collaboration help ensure that designs are informed by practical construction expertise. (3) *Efficiency and Cost-Effectiveness*: Buildability is strongly tied to improved productivity, reduced waste, shortened project durations, and overall cost optimisation. (4) *Safety and Risk Reduction*: Designs that reflect construction realities help mitigate risks, prevent accidents, and promote safe working conditions. (5) *Lifecycle and Sustainability Considerations*: More recent study acknowledges that buildability intersects with sustainability, particularly in reducing resource consumption and supporting efficient use of materials.

Table 3.1, which summarises the definitions provided by the various authors, demonstrates that although terminology and emphasis may differ, the underlying principle remains consistent: buildability is fundamentally about designing in ways that make construction easier, more efficient, and more predictable. Table 3.1 also illustrates the evolution of the buildability concept, from early definitions centred on design clarity, construction empathy, and ease of assembly, to more contemporary interpretations that emphasise the integration of construction knowledge, collaborative decision-making, technological innovation, resource efficiency, safety, sustainability, and broader organisational and procurement considerations.

Table 3.1 Summary of the different definitions of buildability by several authors

No.	*Definition*	*Author(s)*
1	The degree at which the design of a building structure aid construction, subject to the overall requirements for the finished building.	CIRIA (1983)
2	Design and detailing which recognise the construction process in attaining the project goals safely with minimum cost to the client.	Illingworth (1984)
3	Best incorporation of construction experience and knowledge in planning, procurement, engineering and construction to achieve project goals.	CII (1986)
4	Lack of design empathy for construction.	Griffith (1987)
5	Ability to construct a building structure economically, efficiently and to agreed quality levels from its sub-assemblies, components and constituent materials.	Ferguson (1989)
6	The extent to which decisions, made during the whole building procurement process, ultimately facilitate the ease of construction and the quality of the completed project.	Chen et al. (1993)
7	A problem of managing the transfer of appropriate knowledge about the construction process to the design process worker.	Moore (1993)
8	A measure of the expediency with which a building structure can be constructed.	Fisher and Rajan (1996)
9	The degree to which the design of a building is production friendly.	Bamisile (2004)
10	The extent to which a building design facilitates efficient use of construction resources and enhances ease and safety of construction on-site while the client's requirements are met.	Lam et al. (2006)

In synthesising the literature, buildability can therefore be understood as a multidimensional construct that influences project outcomes from planning through to completion. It is both a philosophy and a practical approach that encourages collaboration, proactive design decision-making, and continuous improvement. Ultimately, buildability represents a commitment to designing with construction in mind, ensuring that projects can be delivered efficiently, economically, and safely.

The definition of buildability proposed by the Construction Industry Research and Information Association in 1983 remains the most widely accepted and frequently referenced conceptualisation within construction management literature. CIRIA defines buildability as *"the degree at which the design of a building structure aids construction, subject to the overall requirements for the finished building."* This definition became foundational because it was both concise and comprehensive, capturing the essential relationship between design intent and construction practicality. Moreover, it established buildability not merely as a design characteristic but as a performance measure reflecting how design decisions influence construction outcomes.

Embedded within CIRIA's definition are two important indirect implications. First, buildability is not an absolute condition, but rather a variable one. Projects can demonstrate good, moderate, or poor buildability depending on how effectively the design facilitates construction activities. Later authors echo this interpretation by framing buildability as a spectrum, Fisher and Rajan (1996) describe it as "*a measure of the expediency with which a building structure can be constructed,*" while Bamisile (2004) defines it as "*the degree to which the design of a building is production friendly.*" These definitions reinforce the notion that buildability reflects how supportive or obstructive the design is to efficient building processes.

The second implication of CIRIA's definition is that buildability may occasionally conflict with other project requirements. CIRIA explicitly notes that buildability must be considered *"subject to the overall requirements for the finished building."* This means that architectural ambitions, client specifications, aesthetic considerations, engineering innovations, and heritage constraints may sometimes take precedence over optimal buildability. In such cases, some degree of poor buildability may need to be accepted to satisfy higher-order project demands. Lam et al. (2006) highlight this tension by defining buildability as the extent to which a design supports efficient, safe construction *whilst meeting client requirements.* Thus, achieving perfect buildability is not always possible; instead, designers and constructors must navigate competing priorities.

The broader literature offers complementary perspectives that enrich our understanding of buildability. Illingworth (1984) emphasises the consideration of construction processes in design detailing to achieve safety and cost efficiency. The CII (1986) expands this view by defining buildability, within the US context as constructability, as the optimal integration of construction knowledge into planning, procurement, engineering, and construction. Griffith (1987) frames buildability more critically, describing it as the absence of design empathy for construction, suggesting that poor buildability often arises from neglect of construction realities. Ferguson (1989) focuses on the efficiency with which a structure can be assembled from its components, while Chen et al. (1993) broaden the concept to encompass decision-making throughout the entire procurement process. Moore (1993) conceptualises buildability as a knowledge management issue, specifically the challenge of transferring construction knowledge to design practitioners.

Taken collectively, these definitions illustrate that buildability is a multifaceted concept encompassing design clarity, knowledge integration, resource efficiency, construction safety, and alignment with client objectives. While each author emphasises different aspects, they all converge on the central idea that buildability concerns the ease and efficiency with which construction can be achieved.

In general, therefore, buildability in construction can be understood as a pre-construction evaluative process that reviews building production information, architectural drawings, structural layouts, mechanical and electrical designs, and other project documentation from the viewpoint of constructors. This process aims to ensure that the design is logical, practical, economical, and conducive to safe and efficient construction. By foregrounding construction considerations at the design stage, buildability supports improved project delivery, reduced waste, and enhanced integration across the project team.

After a careful analysis of previous definitions of buildability, the authors of this book define buildability as the extent to which a building's design, enhanced by digital tools and collaborative processes, supports safe, efficient, and sustainable construction through the use of modern methods and technologies while still meeting the intended performance requirements of the completed asset.

3.2 Principles of Buildability

There are a number of fundamental principles that architects and building designers must apply during the design process to achieve good buildability in construction projects. These principles serve as practical guidelines that help ensure the design can be translated into a safe, efficient, and cost-effective construction process. Over the years, different scholars have proposed various sets of buildability principles, often shaped by their research focus, professional background, or industry observations. Although these principles may appear diverse or subjective, they consistently address similar core issues that influence construction performance.

Common themes include access, ensuring that workers, equipment, and materials can reach work areas safely and efficiently; timing, which considers how design decisions influence construction sequencing and scheduling; and skill levels, acknowledging that the availability and capability of labour should align with design expectations. Other recurring principles relate to repetition, which promotes standardisation and reduces complexity; tolerances, ensuring precise interfaces between components; and construction sequences, which help avoid clashes, delays, and unnecessary rework.

Despite variations in terminology or emphasis, the underlying concepts across scholarly contributions remain remarkably consistent. Together, these principles reinforce the idea that thoughtful, practical, and construction-aware design is central to achieving high buildability and improving overall project delivery.

CIRIA (1983) identified seven principles of buildability to include the following:

- Carry out thorough investigation and design
- Plan for essential site production requirements
- Plan for a practical sequence of building operations and early enclosure
- Plan for simplicity of assembly and logical trade sequences
- Detail for maximum repetition and standardisation
- Detail for achievable tolerances
- Specify robust and suitable materials

Adams (1989) proposed three principal criteria for good buildability after analysing the CIRIA's seven principles of buildability. These principal criteria are as follows:

- Simplicity
- Standardisation
- Clear communication

In advancing the management of buildability, the Construction Industry Institute Australia (CIIA) developed 12 key buildability principles that serve as comprehensive criteria for evaluating and improving construction effectiveness. These principles provide a structured framework to guide designers, contractors, and project managers in identifying factors that influence construction ease, efficiency, safety, and quality. They address essential considerations such as early integration of construction knowledge, standardisation, communication, site access, material selection, sequencing, and opportunities for simplification. Collectively, the CIIA principles offer a practical foundation for incorporating buildability throughout the project lifecycle and enhancing overall project performance. These 12 principles are discussed in the following paragraphs.

1 *Integration*: Integration emphasises embedding buildability considerations into the project from the earliest stages of planning. This means treating buildability as a core design and management function, not an afterthought. When integrated effectively, buildability shapes decisions on layout, materials, sequencing, procurement, and stakeholder coordination. It aligns project objectives with constructability requirements and ensures that the design team, contractors, and specialists operate within a unified framework. By integrating buildability into the overall project plan, organisations reduce rework, improve communication, and enhance the predictability of time, cost, and quality outcomes throughout the project lifecycle.
2 *Construction Knowledge*: This principle stresses that practical construction expertise must inform project planning and design. Designers benefit from understanding real-world site conditions, construction methods, labour capabilities, and material limitations. Early contractor involvement, consultations with construction managers, and lessons learned from past projects help identify potential challenges before they materialise on-site. Incorporating construction knowledge leads to design solutions that are more efficient, safer, and easier to deliver. It also ensures that decisions regarding structural systems, detailing, and sequencing align with achievable construction practices and available technologies.
3 *Team Skills*: Buildability depends heavily on the competence and composition of the project team. This principle highlights the need for a balanced team with relevant technical skills, experience, and collaborative ability. A team with inadequate or mismatched expertise increases the risk of impractical designs, poor coordination, and inefficient construction. Selecting specialists early, clarifying roles, and ensuring skill alignment with project complexity are essential. Effective teamwork also facilitates smoother communication between designers, engineers, builders, contractors, and subcontractors, promoting better decision-making and reducing misunderstandings that could compromise construction efficiency.
4 *Business Objectives*: Understanding the client's business model and project objectives is crucial for achieving buildability. When the project team clearly grasps what the client values, whether speed, cost efficiency, flexibility, sustainability, or aesthetic impact, they can make design and construction decisions that support these priorities. This principle ensures that buildability

strategies align with broader organisational goals rather than being applied in isolation. A project becomes more buildable when design intent, performance requirements, and operational needs are harmonised, enabling efficient construction without compromising the client's ultimate vision or functional outcomes.

5 *Available Resources*: This principle stresses the need to match the design with the skills, materials, equipment, and technologies realistically available in the project environment. Overly complex designs requiring specialised labour or rare materials may lead to delays, rework, or increased costs if the necessary resources cannot be secured. By aligning the design solution with available capabilities, whether local labour skills, accessible supply chains, or feasible construction technologies, the project becomes more practical and buildable. This alignment supports efficient delivery, minimises risk, and encourages appropriate innovation.
6 *External Factors*: External factors such as weather, site constraints, regulatory requirements, environmental conditions, market fluctuations, and community impacts can significantly influence construction costs and schedules. This principle stresses the need to anticipate and plan for these influences early in the project lifecycle. Accounting for external factors allows the team to develop resilient designs and construction strategies that minimise disruptions. Effective management of external risks enhances buildability by ensuring logistical clarity, reducing delays, and aligning the project with regulatory or contextual constraints.
7 *Programme*: A realistic, construction-sensitive programme is fundamental to buildability. This principle emphasises developing a project schedule that reflects actual construction sequences, resource availability, and potential constraints. Unrealistic programmes lead to rushed work, increased errors, and higher risk. A well-considered programme ensures logical sequencing, adequate lead times for procurement, and coordinated activities across trades. It also requires commitment from the entire team to avoid deviations that compromise buildability. A properly structured programme enhances efficiency, safety, and cost control.
8 *Construction Methodology*: Design decisions must be aligned with feasible, efficient construction methodologies. This principle ensures that the design supports logical construction processes, whether traditional methods, prefabrication, modular construction, or mechanised techniques. Considering methodology early helps avoid impractical details, allows for streamlined sequencing, and encourages the adoption of efficient technologies. It also ensures that materials, structural systems, and component sizes reflect the chosen method. Ultimately, incorporating methodology into design reduces rework, improves safety, and supports faster, more predictable construction.
9 *Accessibility*: Accessibility refers to providing safe and efficient access for workers, materials, equipment, and maintenance during both construction and operation. Designs should consider movement routes, site logistics, crane positioning, scaffolding, and installation space. Poor accessibility leads to delays,

safety hazards, and higher labour costs. Integrating accessibility considerations early enhances buildability by enabling smoother workflows, reducing handling difficulties, and preventing congestion on-site. It also supports long-term asset management, as accessible designs simplify maintenance and future upgrades.

10 *Specifications*: Specifications must support construction efficiency, clarity, and practicality. Overly complex or ambiguous specifications can lead to misinterpretation, disputes, and rework. This principle encourages designers to select materials, products, and performance criteria that are readily available, compatible with proposed construction methods, and aligned with industry standards. Well-prepared specifications enhance consistency, support quality assurance, and ensure that construction teams understand the project's technical requirements. Clear, buildable specifications reduce risk and facilitate smooth procurement and execution.

11 *Construction Innovation*: This principle recognises that innovative construction techniques, such as modular construction, robotics, digital fabrication, or advanced materials, can significantly enhance buildability. Innovation improves efficiency, quality, and safety while reducing waste and labour intensity. It encourages the project team to explore new technologies and methods that may better suit the project's design and constraints. Adopting innovation must be balanced with feasibility, cost, and resource availability, but when applied appropriately, it can greatly improve buildability and project outcomes.

12 *Feedback*: Feedback involves learning from completed projects through structured post-construction reviews. This principle ensures that successes, failures, and lessons learned inform future designs and construction practices. Effective feedback mechanisms improve buildability over time by identifying recurring issues, highlighting best practices, and supporting continuous improvement. It also strengthens organisational knowledge and enhances collaboration across project teams. When feedback is systematically captured and shared, it becomes a powerful tool for improving future project performance and advancing industry-wide buildability standards.

3.3 Buildability Attributes across Project Phases

Buildability attributes refer to the key characteristics of a design that support the effective integration of construction knowledge throughout the project lifecycle. These attributes help designers, engineers, builders, and contractors align design intentions with practical construction considerations, enabling a balance between project objectives, environmental conditions, and resource constraints. In essence, buildability attributes function as a set of design guidelines or criteria that promote good buildability outcomes. They allow project teams to anticipate potential construction challenges, evaluate alternative solutions, and select design strategies that maximise efficiency, minimise waste, and enhance safety.

A design that consciously incorporates buildability attributes tends to achieve better performance in terms of cost, time, constructability, and sustainability. As such, achieving good buildability requires intentional focus on these attributes from the earliest conceptual stages. Their application ensures that decisions regarding materials, components, construction methods, and sequencing are grounded in practical construction knowledge rather than abstract design preferences. This approach enhances collaboration between designers and constructors, reduces rework, and improves the predictability and overall quality of the construction process.

The following 19 buildability attributes represent a comprehensive set of considerations that designers can apply to improve design buildability. They encompass issues such as simplicity, access, coordination, standardisation, repetition, tolerances, construction logic, and site conditions. Together, these attributes form a structured framework for evaluating the buildability of a proposed design and identifying opportunities for refinement. These buildability attributes are summarised and presented in Table 3.2.

Buildability considerations should begin at the earliest stages of a construction project, progressing in parallel with initial planning, conceptual design, and feasibility studies, and continuing consistently throughout the entire project lifecycle. Integrating buildability from the moment a client's notional idea is formed ensures that construction knowledge informs every major decision, enabling the development of designs that are practical, efficient, and aligned with available resources. Early consideration allows designers to evaluate alternative construction methods, identify potential constraints, and incorporate solutions that minimise complexity, reduce costs, and enhance overall project performance.

For buildability to be effective, it must address each phase of the construction process, from concept development and detailed design to procurement, construction, and eventual handover. This holistic approach ensures that design intentions remain compatible with construction realities such as site conditions, available technology, skill levels, material logistics, and project sequencing. By adopting this perspective, the project team can maximise the use of construction resources, prevent rework, and improve coordination among stakeholders.

Figure 3.1 illustrates the sequential processes or stages at which buildability can be incorporated and highlights the types of construction knowledge required at each stage. This structured view emphasises the continuous and integrated nature of buildability throughout the project.

Figure 3.1 illustrates the dynamic process through which buildability considerations should be integrated across the entire project lifecycle. The diagram highlights how construction knowledge and user knowledge must interact with various project phases, from conceptual planning to eventual use, to support the achievement of project goals.

At the top of Figure 3.1, several contextual influences are shown: socio-political, economic, technical, institutional, and project environment. These represent external and internal constraints that shape decision-making at every stage. Recognising these influences early helps project teams understand risks, regulatory requirements, stakeholder expectations, and the feasibility of design and construction strategies.

Table 3.2 Buildability attributes, descriptions, and considerations

No.	*Buildability Attributes*	*Description*	*Considerations*
1	Consider innovative/ efficient construction techniques	The design solution encourages the use of efficient and innovative construction methods.	Use of Industrialised Building System (IBS). Less in-situ works.
2	Provide complete and clear design information	This attribute considers the coordination of drawings, specifications, updating specifications, and removing ambiguities or misunderstandings.	Coordination of drawings and specifications. Avoidance of missing information.
3	Make best use of prefabrication	Precast or prefabricated self-contained bathrooms and kitchens, etc., reduce the amount of wet trade activities on-site.	Allowing off-site work for typical floor buildings and for non-typical floor buildings. Enabling the adoption of single integrated elements (e.g. whole toilet completed with sanitary ware, piping & finishes) at the discretion of contractor. Optimising the mix of off-site work (e.g. pre-casting) and on-site work (e.g. final levelling).
4	Capitalise on standardisation	Standardisation can be manifested through the repetition of grids, sizes of components, and connection details. It enhanced ease of construction by increasing project performance and reducing the project cost.	Allowing modular layout of components. Allowing a high degree of standardisation and repetition for floor buildings. Allowing the use of standard details with lots of repetitions.
5	Less work below ground	In conditions where the ground is hazardous, poor, or wet, it will slow down the speed and flow of the project. It is crucial to obtain a clear understanding of the conditions likely to be met beforehand. The designer must always bear in mind that work below ground is always difficult to carry out.	Designing for minimum construction time below ground. Designing for safe construction below ground. Considering the effects of below ground work on surrounding buildings, e.g. destabilising foundations.

(*Continued*)

Table 3.2 (Continued)

No.	*Buildability Attributes*	*Description*	*Considerations*
6	Simple detailing	Details should be kept as simple where possible to reduce the learning curve effect of site tradesmen. Reasonable tolerances should be specified.	Designing to aid the visualisation of finished work. Referring to typical/standard details for repetitive items. Using blow-up details to examine possible clashes in the design, e.g. building services clashing with reinforcements.
7	Allow flexibility	The high adaptability of building elements helps save resources and increase the flexibility for change according to actual site conditions being encountered.	Using interchangeable components, e.g. optional left/right orientation of cabinets, sanitary ware, or universal assemblies.
8	Simple installation	Irregular shapes, complex geometrical profiles, complicated installation details, and multidisciplinary designs could burden contractors with additional resources for coordination and site assembly. As such, building designs with simple configurations enable works to be executed in a straightforward manner and facilitate ease of construction. However, this aspect must be balanced with the aesthetic requirements of clients and the artistic aspiration of designers.	Allowing adaptation (e.g. piping around obstacles instead of penetrations) by the contractor on-site without extensive rework. Allowing flexibility in erection/trade sequences (e.g. G/F slab laid after all upper floors). Allowing for early removal of temporary support to leave clear working space.
9	Use visualisation tools	The rapid development of information technology has enabled advanced software to be used for preparing designs that integrate visualisation features. Visualisation has been cited to improve the collaboration between site and design teams in solving buildability problems that may arise during construction.	Enabling design requirements to be easily visualised and coordinated by site staff

(*Continued*)

Table 3.2 (Continued)

No.	*Buildability Attributes*	*Description*	*Considerations*
10	Optimise labour/ skills usage	Design is only good as skills available to execute it, either off-site or on-site. Labour and skills requirements vary between one project and another, between one locality and another. Design must include a realistic assessment of the level of skill likely to be available from appropriately chosen contractors and specialists.	Allowing the use of know-how and labour skills available locally. Allowing economical use of labour (e.g. balancing between labour and plant use). Avoiding as far as possible multiple handling and visits by different trades.
11	Optimise materials usage	Products and materials to be specified during design should be selected with care, particularly those that have not long been established and accepted within the industry. It is recommended that only products and materials, which have been proven suitable to be used, are selected. At the same time, more economical benefits can be gained if local materials are used, as in many cases, materials produced locally are cheaper and easier to get.	Allowing the use of a wide range of materials to fulfil the required performance. Giving rise to lower cutting wastages (e.g. tiles and rebars).
12	Optimise plant and equipment usage	Designing for optimum use of plant and equipment and having the knowledge of them, also designing for temporary plant and equipment anchorage in permanent structures.	Designing for optimum use of plant and equipment. Designing with knowledge of plant and equipment capacities. Allowing the use of local plant and equipment available.
13	Effective site layout	The environment around the construction site has a significant influence on the whole development process. Confined sites possess characteristics concerning the imposed restrictions which substantially hinder the construction process. The smoothness of construction is susceptible to nearby traffic conditions, especially when exits and ingress often encroach upon pedestrian pavements and vehicular routes.	Allowing sufficient working space for labour and plant. Causing less environmental nuisance (e.g. noise, vibration, wastewater, chemical waste, and dust) to surroundings for urban sites. Allowing for early enclosures from weather for high-rise buildings. Allowing for construction traffic on permanent structure early after erection (e.g. left-in steel decking on structural steel).

(*Continued*)

Table 3.2 (Continued)

No.	*Buildability Attributes*	*Description*	*Considerations*
14	Minimise or avoid rework	This attribute will optimise the usage of certain trades by enabling the trade to work with as few reworks as possible.	Avoiding as far as possible multiple visits by different trades.
15	Consider construction sequence	The sequence of installation should not be dictated in the design document but left for the contractor to decide on the sequence for the entire works. For example, ground floor slabs can be constructed before or after superstructure construction to allow flexibility in the timing of underground drainage works.	Letting the contractor decide on the sequence for the entire works.
16	Sufficient tolerance	The tolerances which are normally achievable in site conditions should be properly considered in the design, especially interfacing between different building components such as different products, methods of construction, materials, and method of manufacture. For example, sufficient tolerance needs to be specified to allow the window frame to be fixed onto the window opening without having to do any adjustment or modification.	Specifying tolerances for as many items as possible. Co-ordinating tolerances specifications for interfacing items (e.g. window frame vis-à-vis window opening).
17	Consider the impact of the weather	Considering possible timing to avoid carrying out structural work, external finishes, etc., during rainy days, for high-rise buildings is an important issue.	Considering possible timing to avoid carrying out structural work, external finishes, etc., during the rainy season for high-rise buildings.
18	Consider safety	In high-rise building, allowing safe sequence of trades is an issue. Designers should look into the impact of the design solution on safety of manpower during handling of materials and components, or wherever requirement for access is necessary. The design should be arranged so as to facilitate safe working in areas prone to accidents such as underground and high elevation works.	Allowing safe sequence of trades (e.g. heavy M&E plant hoisted into position before building is fully enclosed) for high-rise buildings. Sizes and weights of materials and components are safe for workers to handle using commonly available plant for high-rise buildings.

(*Continued*)

Table 3.2 (Continued)

No.	*Buildability Attributes*	*Description*	*Considerations*
19	Encourage sustainable construction throughout the project lifecycle	With heightened awareness of environmental pollution, natural resource depletion, and accompanying social problems, sustainable development and sustainable construction have become a growing concern throughout the world.	Design for sustainable materials in the project. Design for method of construction that preserves the environment.

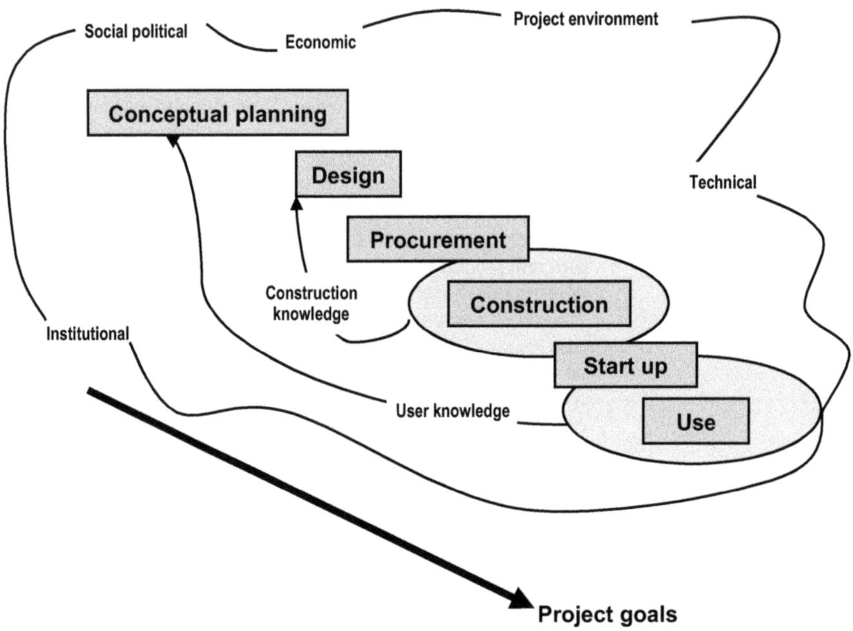

Figure 3.1 Integrated buildability framework across the project lifecycle.

Source: Mydin et al. (2011).

The project lifecycle begins with conceptual planning, where fundamental project objectives, scope, and feasibility are established. At this early stage, buildability can have the greatest impact, as decisions made here significantly influence downstream activities. As the project moves into design, the flow arrows indicate that construction knowledge should be fed directly into the design process. This ensures that proposed solutions remain practical, cost-effective, and aligned with available skills, materials, and technologies.

The procurement stage bridges design and construction. Here, buildability considerations influence contract strategies, material selections, and construction methods. The diagram emphasises that knowledge transfer must continue to flow forward and backward to maintain alignment between design intentions and construction realities.

During construction, both construction knowledge and practical site experience shape the execution of work. The shaded oval around the construction stage highlights its centrality in evaluating whether earlier buildability decisions were appropriate.

As the project transitions to start-up and use, user knowledge becomes increasingly relevant. Lessons learned, operational insights, and feedback on maintainability contribute to continuous improvement and inform future design and construction practices.

Finally, the large arrow pointing towards project goals demonstrates that buildability is not a one-time activity but a continuous, integrated process aimed at enhancing performance, reducing waste, and achieving long-term value throughout the asset's life cycle.

3.4 Factors Affecting Buildability, Problems, and Impacts

3.4.1 Factors Affecting Buildability

There are numerous factors that influence buildability, and understanding these elements is essential for formulating effective strategies to enhance construction efficiency and project outcomes. Improving buildability requires continuous refinement of these factors, particularly during the design phase, where most construction challenges can be anticipated and mitigated. Key considerations include providing contractors with the flexibility to propose alternative construction details, promoting the economical use of labour, materials, and equipment, and ensuring that design requirements are clearly visualised, coordinated, and communicated to site personnel.

Client briefing also plays a critical role, as clear project objectives help designers align functional, aesthetic, and operational needs with construction capabilities. Simplicity in design, such as reducing unnecessary complexity, standardising components, and minimising excessive detailing, can significantly improve buildability by reducing errors, delays, and resource waste. Furthermore, the early involvement of constructors or construction experts fosters better integration of practical knowledge into design solutions, ensuring that potential site constraints, sequencing issues, and resource limitations are addressed before construction begins.

To facilitate structured analysis, these factors have been classified into 16 categories, each representing a key dimension of buildability. These categories and their corresponding factors are summarised in Table 3.3.

Table 3.3 presents a comprehensive classification of the key factors influencing buildability, grouped into 16 categories that collectively shape the efficiency, safety, and quality of construction projects. Table 3.3 begins with site-specific considerations, emphasising the importance of thorough investigations such as soil testing,

Table 3.3 Factors influencing buildability

No.	*Group Factors*	*Factors*
1	Site-specific Factor	Thorough site/ground investigation (e.g. bore holes, topography survey, cable detection, and survey on adjacent buildings).
2	Site Layout, Access and Environment	Allowing sufficient working space for labour and plant. Enabling efficient site layout, storage, and site access Allowing less wet trades on-site Causing less environmental nuisance (e.g. noise, vibration, wastewater, chemical waste, and dust) to surroundings Allowing for early enclosures from weather Allowing for construction traffic on permanent structure early after erection (e.g. left-in steel decking on structural steel)
3	Below Ground	Designing for minimum construction time below ground Designing for safe construction below ground Considering the effects of below ground work on surrounding buildings, e.g. destabilising foundations
4	Co-ordination and Rationalisation of Design Information	Co-ordinating drawings and specifications Updating specifications and removing ambiguities/ misunderstandings Dimensional co-ordination Providing/facilitating combined services drawings Showing accurate positions for pipe sleeves and penetrations
5	Detailing	Specifying tolerances for as many items as possible Co-ordinating tolerances specifications for interfacing items (e.g. window frame vis-à-vis window opening) Designing to aid visualisation of finished work Referring to typical/standard details for repetitive items Using blow-up details to examine possible clashes in the design, e.g. building services clashing with reinforcements
6	Flexibility	Designing for interchangeability (e.g. left/right orientation of fittings, such as cabinets, and kitchen sinks) and sub-assemblies
7	Tools, Plant, and Equipment	Designing for optimum use of plant and equipment Designing with knowledge of plant and equipment capacities Designing for temporary plant and equipment anchorages in permanent structure
8	Materials, Fittings, Products and Sub-assemblies	Designing for locally available materials/fittings/ products/sub-assemblies (including imports) When imported materials/fittings/products/sub-assemblies are specified, consider supply conditions (e.g. checking lead-times and foreseeable shortages) Specifying robust and suitable materials/components or giving directions for protecting fragile items (e.g. precast stairs) Designing to facilitate the care and protection of completed works by contractors

(*Continued*)

Table 3.3 (Continued)

No.	*Group Factors*	*Factors*
9	Use of Resources	Allowing the use of plant and equipment available locally Allowing the use of know-how and labour skills available locally Allowing economical use of labour and plant (e.g. balancing between labour and plant use to reduce overall cost) Avoiding as far as possible multiple handling and visits by different trades
10	Material Systems	Allowing the use of wide range of materials to fulfil the required performance Giving rise to lower cutting wastages (e.g. tiles and rebars)
11	Installation	Allowing easy connection/interfacing between components Allowing adaptation (e.g. piping around obstacles instead of penetrations) by contractor on-site without extensive rework Specified tolerances capable of being achieved Allowing easy installation without complicated fixings Allowing flexibility in erection/trade sequences (e.g. G/F slab laid after all upper floors) Allowing for early removal of temporary support to leave clear working space
12	Standardisation	Uncomplicated geometry, layout, and shape Allowing the modular layout of components Allowing a high degree of standardisation and repetition Allowing the use of standard details with lots of repetitions, thereby facilitating the learning curve of workers to be built up fast
13	Prefabrication	Allowing prefabrication off-site Enabling the adoption of single integrated elements (e.g. whole toilet completed with sanitary ware, piping & finishes) at the discretion of contractor Optimising the mix of off-site work (e.g. prefabrication, pre-casting, and preassembly) and on-site work (e.g. final levelling and fixing)
14	Innovations	Designing to allow for innovative construction techniques to be proposed by contractor Suggesting non-obligatory construction methods for contractor to consider
15	Weather	Considering possible timing to avoid carrying out structural work, external finishes, etc., during rainy/typhoon season
16	Safety	Allowing safe sequence of trades (e.g. heavy M&E plant hoisted into position before the building is fully enclosed) Sizes and weights of materials and components are safe for workers to handle using commonly available plant

topographic studies, and assessments of adjacent structures. These early insights help prevent costly surprises during construction. Subsequent factors highlight the significance of site layout, access, and environmental conditions, where adequate working space, efficient storage, and reduced wet trades contribute to smoother operations. Environmental concerns, such as noise, dust, and vibration, underscore the growing emphasis on sustainable and community-sensitive construction. Other groups, such as below ground work, coordination of design information, and detailed specifications, stress the importance of accurate documentation, dimensional consistency, clash avoidance, and clear communication between design and construction teams.

The remaining categories focus on optimising construction practicality and resource efficiency. Flexibility in design, effective use of tools and equipment, and the selection of appropriate materials all contribute to improved productivity and reduced waste. Table 3.3 also highlights the importance of installation efficiency, standardisation, and prefabrication, which enhance speed and consistency while reducing on-site labour intensity. Innovation is encouraged through designs that allow contractors to suggest modern construction techniques. Weather-related considerations and safety-focused planning ensure that construction activities remain resilient and secure under varying conditions. Overall, Table 3.3 underscores that buildability is a multidimensional concept, requiring holistic integration of technical, environmental, logistical, and human factors across the entire project lifecycle.

3.4.2 Causes of Buildability Problems

The construction industry is a vital sector of any national economy. In both developed and developing countries, it is regarded as a part of the economy that transforms various resources into constructed facilities through planning, design, and construction. The industry is often considered a key driver of economic development, particularly in developing countries. A French saying underscores its importance: "*When the construction industry prospers, everything prospers*." This reflects the reality that other sectors of the economy rely, directly or indirectly, on the products and services provided by construction to operate effectively.

The industry plays a critical role in the development and maintenance of a nation's physical infrastructure, supporting other sectors of the economy and contributing significantly to the overall gross domestic product (GDP). Its importance is commonly measured by its economic impact, such as its contribution to GDP and the employment it generates. Globally, the construction industry employs approximately 7% of the workforce and spends around $10 trillion annually on construction-related goods and services, equivalent to 13% of the world's GDP, making it one of the largest sectors in the global economy.

The construction industry is dynamic and multifaceted, involving numerous participants, including clients, consultants, contractors, regulators, stakeholders, artisans, and labourers. While the industry is crucial to national development, it also faces significant challenges, including buildability problems. These issues often arise from factors such as designing without contractor input, disregarding contractor-proposed changes, and tight timelines for design and tendering, which limit the opportunity for careful planning and accurate cost estimation. Additionally, the highly fragmented nature of project teams, with specialised consultants handling different aspects, complicates accountability for implementing buildability improvements.

The causes of buildability problems can be grouped into three main categories, as shown in Table 3.4. Understanding these causes is essential for enhancing construction project performance and achieving more efficient and effective project delivery.

Table 3.4 Summary of causes of buildability problems

No	*Category*	*Causes of Buildability Problems*
1	Inception	Inadequate project briefing and poor scoping Budget based on incomplete data and fragmented decision-making Poor estimating, functionality, viability, and defect analysis Lack of pre-design meeting Unavailability of functionality and maintainability statistics and data Economic constraints Lack of construction experience by the clients Resistance of the client to buildability programme Lack of financial incentive for design Highly fragmented roles and specialisation of various consultants in a project team
2	Design stage	Insufficient data and drawing information Non-standardisation Poor specification and choice of materials Inaccurate drawing scale, design, and detailing Lack of builder's input into design Fragmented design process Lack of awareness of buildability concept Lack of design quality management Under design and over design Ignorance of contractors' proposed changes Omission and discrepancies in the drawing Lack of design rationalisation Poor tolerance accommodation Tight timeframe for designing and tendering Choice of poor construction methods Designers' poor knowledge and experience in construction Misinterpretation of clients' needs and requirements
3	Construction stage	Poor construction methodology Communication breakdown Incompetent and lack of skilled and technical personnel Poor workmanship and control system Poor information and control system Omission and divergence No document of lessons learnt Lack of adequate qualification of contractor Adversarial working relationship/environmental Poor monitoring of work progress

Table 3.4 provides a comprehensive overview of the causes of buildability problems, categorised across the inception, design, and construction stages of a project. Understanding these causes is crucial for improving project performance, enhancing efficiency, and minimising delays, cost overruns, and defects. At the inception stage, buildability problems primarily arise from inadequate project planning, poor scoping, and limited client knowledge. Incomplete project briefings, budgets based on fragmented data, and the absence of pre-design meetings can lead to unrealistic expectations and misaligned project objectives. The lack of functionality and maintainability data, combined with economic constraints, further complicates project planning. Additionally, limited client experience in construction, resistance to buildability programmes, and the absence of financial incentives for improved design hinder early consideration of constructability issues. The highly fragmented roles of various consultants exacerbate these problems, reducing accountability and impeding integrated planning. Deficiencies at this stage often propagate into later phases, increasing the likelihood of buildability challenges throughout the project lifecycle.

During the design stage, buildability issues frequently emerge from insufficient data, incomplete drawings, poor specification of materials, and non-standardised processes. A fragmented design approach, coupled with the lack of builder input, can result in designs that are difficult or inefficient to construct. Additional contributing factors include under- or over-design, omission of critical details, poor tolerance accommodation, tight timelines for designing and tendering, and designers' limited construction experience. Misinterpretation of client requirements and ignoring contractor-proposed changes further exacerbate these problems. The absence of design quality management and inadequate awareness of buildability concepts can lead to designs that are theoretically sound but impractical in execution. Addressing these issues requires early contractor involvement, thorough design review, and integrated design processes that prioritise constructability from the outset.

In the construction stage, buildability problems tend to manifest as execution challenges, often caused by inadequate construction methods, poor workmanship, and a shortage of skilled personnel. Communication breakdowns, insufficient monitoring, and the lack of documented lessons learned further exacerbate these challenges. Omissions, divergences from design, adversarial working relationships, and insufficient contractor qualifications also contribute to inefficiencies and defects. Effective construction management, skilled supervision, proper documentation, and a collaborative working environment are essential for mitigating these issues and ensuring that designs are successfully translated into physical structures.

Overall, buildability problems are interconnected across all project stages. Weaknesses in the inception phase, such as poor planning and fragmented roles, often carry over into the design phase, resulting in impractical or overly complex designs. These, in turn, create challenges during construction, where execution may deviate from design due to poor methods, insufficient skills, or ineffective communication. A holistic approach is therefore necessary to address buildability problems, including early contractor involvement, improved stakeholder communication, adequate allocation of resources, skilled personnel, and robust design and construction quality management systems. By systematically addressing the causes at each stage, project performance can be enhanced, minimising delays, cost overruns, and quality defects while improving overall constructability.

3.4.3 Impact of Buildability Problem on Construction Projects

Buildability problems have significant negative implications for construction projects and the industry as a whole. Rather than contributing to project success, these problems often result in delays, cost overruns, compromised quality, and inefficient use of resources. Such outcomes not only affect individual projects but also damage the overall image and credibility of the construction industry. Clients, stakeholders, and end-users may lose confidence in the ability of construction firms to deliver functional, durable, and cost-effective buildings. Therefore, understanding the impact of buildability problems is critical, as it provides the foundation for developing strategies to mitigate these issues and enhance constructability in future projects.

Research studies have highlighted several specific impacts of buildability problems. One major concern is the inability of clients to receive value for the money invested, which can undermine trust and affect the financial viability of projects. Other impacts include poor serviceability and functionality of constructed facilities, structural instability, frequent component failures, and increased risk in building usage. These outcomes not only lead to additional repair and maintenance costs but also reduce the lifespan and performance of buildings. Furthermore, recurring buildability issues can contribute to inefficiencies in project delivery and exacerbate adversarial relationships among stakeholders.

A review of the literature identifies 16 key impact factors associated with buildability problems. To facilitate systematic study and practical application, these impact factors are often grouped into three main categories, as summarised in Table 3.5. Categorisation helps in understanding the nature and scope of the problems, allowing project teams to prioritise interventions and implement effective measures that enhance buildability. By addressing these impact factors proactively, construction professionals can improve project outcomes, optimise resource use, ensure better quality control, and enhance client satisfaction, thereby strengthening the overall performance and reputation of the construction sector.

Table 3.5 highlights the multifaceted impacts of buildability problems on construction projects, demonstrating how these issues affect clients, the lifecycle of buildings, and sustainability. One major area of concern is the failure to meet client requirements. Buildability problems can prevent clients from receiving value for money, as designs that are difficult to interpret or construct often result in inefficiencies and additional costs. Inconsistencies in construction, idleness, wastage of resources, and lax quality control are all common consequences of poor buildability, further compromising project outcomes. When client expectations are not met, it undermines confidence in the construction industry and diminishes the perceived professionalism and reliability of the project team.

Table 3.5 Impact of buildability problems on construction project

No	*Main Category*	*Buildability Problems Impact Factors*
1	Client requirement	Inability to give client value for money Difficulty in constructing the design and working drawing to details Inconsistency in construction Lack of efficiency, idleness, and wastage of construction resources Quality control and management laxity
2	Buildability lifecycle and maintenance	Poor serviceability, functionality, and structural instability as a result of reconstruction Excessive cost and probable abandoning of project Delay due to uncertainty in design and interpretation Excessive variation and disorganisation Impossible cost control and management failure
3	Sustainability	Possible building collapse, component failure, and high risk in usage Dimensional inaccuracies and high defect rate Unpredictable life cycle and deterioration pattern of component and subsystem Unsustainable building component Constant troubleshooting Poor building construction form.

The lifecycle and maintenance of constructed facilities are also severely affected by buildability problems. Poor serviceability, compromised functionality, and structural instability often arise from reconstruction or remedial works required to address initial design and construction flaws. Excessive project costs, potential project abandonment, and delays caused by uncertainties in design interpretation are common outcomes. Furthermore, excessive variations, disorganisation, and challenges in cost control and project management can impede timely project delivery. These impacts highlight how early-stage design and construction decisions directly influence the operational effectiveness and long-term viability of a building.

Sustainability is another critical dimension affected by buildability problems. Buildings with dimensional inaccuracies, high defect rates, and unpredictable deterioration patterns compromise structural reliability and user safety. In extreme cases, component failure or even building collapse may occur, posing a high risk to occupants. Additionally, unsustainable components, constant troubleshooting requirements, and poor construction form exacerbate long-term maintenance challenges, reducing the overall lifespan and efficiency of the facility. Addressing these sustainability-related impacts requires careful attention to constructability during design and construction, ensuring that building components are durable, reliable, and maintainable over time.

Overall, the impacts of buildability problems are interconnected, influencing client satisfaction, operational performance, cost efficiency, and sustainability. By systematically understanding these impact factors, construction professionals can develop strategies to mitigate risks, enhance buildability, and improve the overall quality and reliability of construction projects.

3.5 Strategies for Improving Buildability

The construction industry plays a vital role in national development, contributing significantly to economic growth, employment, and the creation of physical infrastructure. Despite its importance, the industry continues to suffer from persistent performance challenges, many of which are linked to poor buildability. Buildability problems often manifest as cost overruns, delays, rework, construction disputes, safety issues, and reduced productivity. These issues arise from a variety of root causes, including non-standardisation of design, inadequate early-stage integration of construction expertise, and a general lack of awareness or appreciation of buildability principles among project stakeholders. Designers may overlook construction practicalities, either due to limited site experience or insufficient collaboration with builders. Similarly, fragmented industry structures and weak communication channels between design and construction teams contribute to poor coordination, resulting in deficient detailing, unclear working drawings, and design decisions that are difficult, time-consuming, or costly to implement.

Other factors exacerbating buildability challenges include strained relationships between designers and contractors, reluctance by clients to invest in buildability enhancement initiatives, and a widespread hesitance to allocate additional time and resources during the conceptual and design phases. This short-term mindset often leads to long-term inefficiencies during construction. Moreover, inadequate consideration of modern construction techniques, site constraints, material availability, and user needs can further limit a design's constructability. Collectively, these issues undermine project performance and hinder the industry's ability to deliver high-quality, sustainable, and cost-effective built assets.

Buildability, however, offers a pathway to overcoming many of these entrenched challenges. When properly applied, buildability enhances project outcomes by promoting clarity, simplifying construction processes, reducing waste, improving safety, and facilitating seamless coordination among project participants. Designers play a crucial role in this process, as their decisions fundamentally shape the ease or difficulty with which a building is constructed. Recognising the importance of their role, this section of this book introduces a set of strategies aimed at improving the buildability of building designs. These strategies, summarised in Table 3.6, serve as practical guidelines to help designers integrate construction knowledge, anticipate site realities, and optimise design solutions.

A thorough understanding and implementation of these strategies ensure that building designs align more closely with construction capabilities and constraints. For instance, adopting standardised components, improving detailing, engaging contractors early, and utilising modular or prefabricated systems can significantly enhance buildability. Similarly, clear communication, coordination of design documents, and the selection of materials that are easy to install and locally available

Table 3.6 Strategies for improving buildability of building designs

No.	*Strategies*
1	The benefits of improved buildability should be made known to designers.
2	More technical literature for improving buildability should be provided to designers.
3	More education and training on buildability should be given to designers.
4	Designers should gain more experience with construction operations.
5	More time should be allowed for designers to consider buildability issues.
6	Clear designation of responsibility for improving buildability is required.
7	Designs should be checked for buildability before issuing for tender.
8	Design team should share the costs of claims arising from design coordination problems (e.g. building services clash with architectural or structural details).
9	The use of alternative procurement approaches enabling early contractor participation in the design stage should be promoted.
10	Formal channels for co-ordination and communication between constructors and designers should be set up for improving buildability.
11	Clients should provide clear requirements on buildability, apart from time/cost/quality targets.
12	Technical guidelines for improving buildability should be provided by professional bodies.
13	Statutory requirement of achieving minimum buildability results should be enforced.
14	Incentives (e.g. awards) should be rewarded for good buildability of designs.
15	A benchmarking system should be established on buildability results.

contribute to smoother construction processes. Design appraisal that incorporates buildability considerations ultimately leads to buildings that perform better not only for constructors but also for users and the environment. By embedding these principles at every stage of the design process, the construction industry can achieve greater efficiency, higher quality outcomes, and improved sustainability.

Table 3.6 presents a comprehensive set of strategies aimed at enhancing the buildability of building designs, addressing both the technical and organisational challenges that commonly undermine construction performance. The first group of strategies (Items 1–5) focuses on strengthening designers' awareness, knowledge, and competence in buildability. By ensuring that designers understand the benefits of buildable designs, such as reduced construction costs, improved safety, and faster project delivery, they are more likely to integrate these principles into their work. Providing access to technical literature, continuous training, and opportunities to gain hands-on construction experience further empowers designers to make informed decisions that align with site realities. Additionally, allocating sufficient time during the design phase allows for deeper scrutiny of constructability issues, mitigating errors that arise from rushed or incomplete designs.

The second group of strategies (Items 6–15) addresses structural and organisational measures needed to embed buildability into project delivery systems. Clear designation of responsibility ensures that buildability is not treated as an afterthought but as a managed process with accountability. Buildability checks before tendering, cost-sharing for coordination-related claims, and early contractor involvement all promote collaborative design development and minimise downstream problems. Establishing formal communication channels between designers and constructors fosters knowledge exchange and helps resolve technical issues early. Clients also play a critical role by articulating explicit buildability expectations and supporting frameworks that promote high-quality design outcomes. Professional bodies can reinforce these efforts by publishing guidelines, while regulatory authorities may enforce minimum buildability standards. Incentive schemes and benchmarking systems further encourage a culture of continuous improvement, recognising excellence and providing measurable performance indicators. Collectively, these strategies form a holistic approach to elevating buildability across the construction industry, leading to better project outcomes and more sustainable practices.

3.6 Buildability in Greenfield and Brownfield Projects

Buildability plays a crucial role in determining the efficiency, cost-effectiveness, safety, and sustainability of construction projects. While buildability principles apply to all forms of development, their practical implications differ significantly in greenfield and brownfield project environments. Greenfield projects involve construction on undeveloped land with minimal constraints, whereas brownfield projects concern sites that have been previously developed, used, or contaminated. Because the characteristics, risks, and conditions of these two project types differ, the strategies required to optimise buildability must also be tailored accordingly.

Understanding buildability in both contexts is increasingly important as global urbanisation, sustainability concerns, regulatory pressures, and land scarcity shape construction practices. Cities are witnessing a rise in redevelopment efforts, while expansion into new territories continues. This duality means that design teams, planners, and contractors must understand how buildability principles can be applied and adapted to maximise project outcomes in both greenfield and brownfield scenarios.

3.6.1 Understanding Buildability in Greenfield Contexts

Greenfield projects refer to developments carried out on land that has not been previously built upon or significantly altered by human activity. From a buildability perspective, these sites present a unique advantage: they offer a blank canvas for construction teams to plan, design, and execute works with minimal physical limitations. The absence of existing structures, underground utilities, and residual contaminants greatly enhances the ease with which designers and contractors can conceptualise and implement buildable solutions. This freedom allows project teams to prioritise efficiency, constructability, and long-term performance right from the early planning stages.

One of the most significant buildability advantages of Greenfield sites is the flexibility they offer in terms of spatial planning and layout optimisation. Designers can strategically arrange building footprints, access routes, material storage zones, and operational areas without having to navigate around pre-existing constraints. This flexibility improves site logistics, reduces unnecessary material handling, and minimises construction clashes. The ability to plan wide and unobstructed access routes, for example, supports the movement of heavy machinery, prefabricated components, and bulk materials. Such efficient logistics contribute directly to improved buildability by shortening construction times, reducing labour intensity, and lowering operational risks.

Furthermore, Greenfield sites create opportunities for the seamless integration of modern construction technologies. Techniques such as modular construction, off-site fabrication, robotics, and automated systems often require ample space for component assembly, storage, and crane operations. Greenfield conditions enable the adoption of these methods without spatial compromises, leading to higher precision, improved quality control, and reduced on-site activities. The ability to incorporate modularisation from the design stage also means that buildings can be engineered for simplified assembly, fewer on-site connections, and enhanced structural coordination, key factors that improve buildability outcomes.

Environmental and geotechnical conditions further influence buildability in Greenfield projects. In many cases, these sites offer favourable soil profiles, stable ground conditions, and predictable topography, which reduce the complexity of foundation works. While some Greenfield sites may still present environmental sensitivities, such as protected flora or fauna, these considerations are typically easier to manage compared to the risks associated with brownfield contamination. Predictable site conditions enable designers and engineers to select foundation systems, structural forms, and material choices that enhance constructability while minimising uncertainty and costly redesigns.

Regulatory environments also tend to be more straightforward in Greenfield developments. Since the land has not been previously built upon, issues related to building reuse, heritage conservation, or the disturbance of existing services are greatly reduced. This simplifies approval processes and provides project teams with more time and clarity to incorporate buildability principles into designs. Early collaboration between designers, clients, and contractors is often more effective in such contexts, as decision-making is not constrained by legacy site conditions.

Overall, Greenfield projects provide ideal conditions for achieving high levels of buildability. Their inherent flexibility allows for the optimisation of design, logistics, technology adoption, and construction sequencing. By leveraging these advantages, project teams can deliver more efficient, cost-effective, and resilient building outcomes while reducing construction risks and enhancing project performance.

3.6.2 Understanding Buildability in Brownfield Contexts

Brownfield projects refer to developments carried out on land that has been previously used, built upon, or otherwise altered by earlier human activity. From a buildability perspective, these projects present a complex set of challenges that demand advanced planning, multidisciplinary coordination, and adaptive construction strategies. Unlike Greenfield developments, Brownfield sites are rarely blank canvases. They may contain old buildings, deteriorated structures, underground utilities, redundant foundations, or remnants of industrial facilities. Each of these elements introduces additional layers of difficulty that directly affect buildability. Consequently, successful execution in such contexts relies on the ability of designers and contractors to integrate thorough site assessments with practical construction solutions.

One of the most significant buildability concerns in Brownfield projects is the presence of unknown or poorly documented underground conditions. These may include old service lines, abandoned pipes, obsolete electrical cables, or concealed basements. Encountering such elements unexpectedly can lead to design changes, delays, cost overruns, and increased safety risks. Therefore, extensive preliminary investigations, such as geotechnical surveys, utility mapping, structural assessments, and environmental audits, are critical for informing buildable designs. These investigations reduce uncertainty and allow project teams to design foundations, structural systems, and service routes that align with actual site conditions.

Environmental contamination represents another major constraint with direct implications for buildability. Many Brownfield sites, particularly former industrial or commercial zones, contain pollutants such as heavy metals, hydrocarbons, or asbestos. Remediation efforts often require specialised processes, strict regulatory compliance, and careful sequencing of construction activities. Designers must integrate remediation strategies into the construction programme by allocating adequate time, space, and budgetary provisions. This may include designing temporary works, isolation zones, access control routes, and material handling procedures that ensure safe and compliant execution.

In addition to environmental and subsurface challenges, existing structures, whether derelict or still functional, frequently impose spatial and logistical constraints. Working within tight or confined urban Brownfield settings limits access for machinery, restricts material movement, and complicates the layout of temporary facilities. Buildability in such contexts requires meticulous planning of site logistics, crane positioning, modular assembly strategies, and phased construction sequencing. Adaptive design approaches, such as designing smaller prefabricated elements or reconfiguring construction zones, enhance constructability under these constraints.

Heritage preservation and community sensitivities also play a significant role in shaping buildability in Brownfield projects. Some sites may contain historically significant buildings or culturally important features that must be retained or integrated into new developments. This limits the degree of design flexibility and often necessitates specialised construction methods to avoid structural damage. Likewise, Brownfield sites in densely populated areas require careful noise, dust, and vibration control. These considerations influence choices of materials, methods, and equipment, and may restrict working hours or require silent piling, dust suppression, or low-vibration techniques.

Finally, legal and regulatory complexities often accompany Brownfield developments. Ownership disputes, easements, or obligations tied to previous uses can delay approvals and restrict design options. To maintain buildability, project teams must engage legal experts early, ensure design compliance with regulatory expectations, and incorporate buffer time for approvals. Effective communication among designers, contractors, clients, regulators, and local communities is essential for aligning expectations and preventing coordination conflicts during construction.

Overall, Brownfield projects demand a far more sophisticated approach to buildability than Greenfield developments. By embracing detailed investigations, adaptive design, risk-informed planning, and collaborative execution, project teams can overcome the inherent complexities and transform previously used land into functional, sustainable, and buildable developments.

3.7 Theoretical Perspectives on Buildability

Theories serve as structured frameworks that help scholars and practitioners predict, explain, and interpret phenomena, enabling the expansion, refinement, and sometimes the challenge of existing knowledge. A theoretical perspective refers to a coherent set of assumptions, concepts, and propositions that provide a lens through which a particular issue or phenomenon is understood. Unlike practical explanations, theoretical interpretations rely on abstract principles and conceptual reasoning to explain why things occur in a certain way. Within the context of buildability, theoretical perspectives offer deeper insight into the underlying mechanisms that shape design decisions, construction efficiency, stakeholder behaviour, and project performance.

This section of this book provides the theoretical foundations that support buildability principles, highlighting how established theories from various disciplines illuminate the logic behind improved design and construction integration. By grounding buildability in theory, the discussion moves beyond practical recommendations to present a conceptual justification for why certain strategies enhance constructability and project outcomes.

Several mainstream theories are particularly relevant in explaining buildability. Rationalisation theory underscores the importance of reducing complexity and streamlining processes to achieve greater efficiency. Simplicity theory and standardisation theory similarly advocate for straightforward designs and repeatable elements, which directly enhance constructability. Systems theory provides a holistic view of the construction process, emphasising interdependence, coordination, and integration among project components. Meanwhile, signalling theory and agency theory address behavioural dimensions, such as communication, incentives, and information asymmetry between clients, designers, builders, and contractors. Transaction cost economics (TCE) further explains how buildability decisions can minimise costs associated with coordination, uncertainty, and contractual relationships. Together, these theories form a robust conceptual foundation for understanding and advancing buildability principles.

3.7.1 Rationalisation Theory

The theory of rationalisation was first acknowledged in the work of Ernest Jones in 1908, where he described rationalisation as the process of creating logical, alternative explanations to justify thoughts, decisions, or actions. In its simplest form, rationalisation refers to applying logical reasoning to achieve better outcomes.

Within industrial contexts, rationalisation has evolved into a systematic approach that encompasses methods and techniques aimed at eliminating waste and inefficiencies in areas such as time, labour, materials, processes, and tools. By reducing unnecessary activities and improving workflow organisation, rationalisation lowers production costs, increases efficiency, and ensures that market demands are met more effectively.

In construction, the theory of rationalisation proposes that project performance can be significantly improved when resources are systematically organised, processes are simplified, and waste is minimised. This theory emphasises not only the elimination of inefficiencies but also the creation of alternative strategies or explanations that allow project teams to justify decisions leading to superior results. Research grounded in the theory of rationalisation suggests that when participants in a project or organisation develop coherent, reasonable alternative explanations for actions, they tend to choose methods that enhance performance. In this sense, rationalisation becomes an active tool for improvement, shaping how individuals and groups understand challenges and identify solutions.

The relevance of rationalisation theory to buildability lies in its focus on process optimisation. Buildability is fundamentally concerned with designing and organising construction works in ways that simplify execution, reduce errors, minimise waste, and improve overall project outcomes. Rationalisation theory supports these goals by providing a conceptual basis for identifying inefficiencies in design and construction and developing logical strategies to overcome them. It underscores that improved buildability is not achieved by chance but through deliberate, reasoned decision-making grounded in efficiency and clarity.

Beyond its application to construction management, rationalisation theory has broader uses in understanding human behaviour and organisational decision-making. It helps reveal agents' motivations, interpret experimental evidence where observed choices may not align with stated preferences, and uncover hidden tendencies that influence actions. Rationalisation theory also aids in explaining behavioural shifts even when incentives or circumstances remain unchanged. Within game theory, it is used to evaluate how individuals justify decisions under strategic interactions. These wide-ranging applications highlight its conceptual strength and reinforce its suitability for supporting buildability analysis.

In the context of building construction, rationalisation theory enhances the rigor of buildability assessment by advocating for structured, efficient, and well-justified planning. It aligns closely with the S.M.A.R.T. approach, Specific, Measurable, Achievable, Realistic, and Time-Bound, ensuring that buildability goals are clearly defined, practical, and results-oriented. When rationalisation principles guide buildability planning, decisions become more transparent, alternatives are evaluated more critically, and performance improvement becomes more achievable.

Conversely, without grounding buildability assessment in rationalisation theory, improvements may be inconsistent or difficult to sustain, as decisions lack the logical structure necessary for identifying and reducing inefficiency. Therefore, rationalisation theory plays a critical role in guiding buildability research and practice, offering a strong conceptual foundation for enhancing construction efficiency and performance.

3.7.2 Simplicity Theory

The theory of simplicity was first introduced by S. Shelah in the early 1980s as part of model theory, but its conceptual relevance has since extended into various fields, including design, engineering, and construction. At its core, simplicity theory aims to explain phenomena in clear, straightforward ways by reducing complexity where possible and ensuring that ideas, processes, and systems remain understandable. Rather than defining simplicity in absolute terms, the theory frames it as the relative difference between expectation and observation; something is considered simple when it aligns with or is easier than what is anticipated.

Simplicity theory recognises two principal forms of simplicity: ontological simplicity and syntactical (structural) simplicity. Ontological simplicity is rooted in the philosophical principle of Occam's razor, which argues for parsimony, avoiding unnecessary assumptions, components, or complications. In essence, the simplest explanation or design is often preferred, provided it effectively serves its purpose. Syntactical simplicity, on the other hand, focuses on the organisation and relationships between elements in a design or system. It concerns the number of components involved, the complexity of their interactions, and the clarity of the patterns that emerge from their arrangement. Together, these forms highlight the need for minimalism in both content and structure.

Simplicity theory provides explanations for several phenomena that influence interest, efficiency, and decision-making, especially within construction. For instance, building designs that are well-detailed and easy to interpret tend to be perceived as more elegant, functional, and appealing. Construction workers are more likely to adopt a particular method when it is simple, logical, and efficient. Builders often value materials or components that are remarkable due to their clarity and simplicity of use. Similarly, contractors are far more able to deliver projects on time and within budget when designs avoid unnecessary complexities, reduce ambiguities, and streamline construction processes.

Buildability research indicates that simplicity theory plays an important role in shaping the decisions made by building designers. When simplicity is absent, when drawings are complicated, unclear, or overloaded with unnecessary details, mistakes on-site increase, coordination becomes more difficult, construction time extends, and costs escalate. Poor project performance often persists under such conditions, particularly in environments where site operatives rely heavily on design clarity to execute construction activities safely and efficiently.

Integrating simplicity theory into buildability assessment ensures that building designs are straightforward, logical, and construction-friendly. A modelled design guided by simplicity reduces errors, enhances communication across disciplines, and supports more predictable construction outcomes. It also allows contractors to visualise construction sequences more easily, facilitates smoother installation of services, and minimises costly revisions during execution. Furthermore, simplicity supports standardisation and rationalisation—two stronger pillars of good buildability practice.

Therefore, the consideration of simplicity theory during the design and buildability assessment phase is essential. By eliminating unnecessary complexity and providing clear, intelligible design information, project teams can significantly improve efficiency, safety, and overall project performance. Simplicity not only strengthens buildability but also ultimately contributes to a more reliable, cost-effective, and sustainable construction process.

3.7.3 Standardisation Theory

Standardisation is founded on the principle of collective agreement. A standard emerges when two or more stakeholders reach consensus on the best or most acceptable way of performing an activity, producing a product, or defining a process. This consensus-based approach makes standardisation a collaborative and negotiated process rather than a rigid technical prescription. It involves input from diverse parties such as firms, industry practitioners, users, interest groups, standards-setting bodies, regulatory agencies, and governments. In the construction industry, the establishment of standards requires alignment between architects, engineers, builders, contractors, suppliers, regulatory authorities, and sometimes clients. Only when mutual agreement is achieved can a standard be formally endorsed and implemented.

Standardisation is defined as the development and implementation of technical standards that reflect the agreements reached by participating stakeholders. The process typically involves several interconnected activities, including design, sense-making, deliberation, testing, negotiation, and validation. These activities collectively lead to what is often referred to as "*technical closure*," meaning that the final standard represents a stabilised solution that resolves differences and provides clear guidance for practice. Figure 3.2 summarises these standardisation activities, illustrating how stakeholder inputs converge into a unified technical specification.

In the construction sector, standardisation plays a particularly crucial role because construction projects involve multiple interdependent systems, large teams, and a wide range of materials, methods, and technologies. By establishing uniform standards for design, materials, installation procedures, documentation formats, safety requirements, and quality parameters, organisations can reduce variability and improve predictability in project outcomes. Standards help ensure that products and processes are consistent across different project teams, contractors, and geographical locations. This consistency enhances interoperability, simplifies communication, and reduces the likelihood of errors, omissions, or conflicts during construction.

Standardisation is also a powerful tool for continuous improvement. When standards are properly implemented and regularly updated, they create structured pathways for capturing lessons learned, integrating new technologies, and embedding better practices into ongoing and future projects. In this context, standards become dynamic organisational assets rather than static documents. The application of standardisation in construction can significantly minimise defects in building projects or even eliminate them altogether. For example, standardised detailing, prefabricated components, modular design, and uniform installation methods contribute to improved build quality, reduced rework, and shorter construction durations.

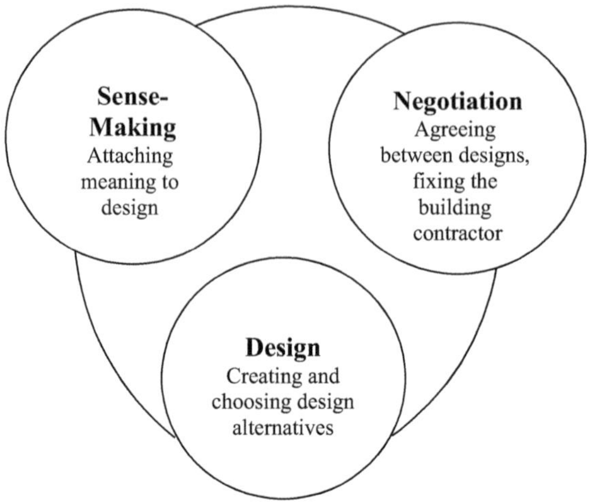

Figure 3.2 Standardisation activities.

Buildability research consistently indicates that standardisation has a positive impact on construction performance. Standardised designs are easier to interpret, coordinate, and construct. Workers become more efficient due to familiarity and repetition, reducing skill-related errors and increasing productivity. Standardised materials and components enhance supply chain efficiency and reduce long lead times, while standardised documentation promotes better communication among project participants. These benefits collectively contribute to enhanced buildability.

In essence, for buildability objectives to be fully realised, standardisation must underpin the design and construction process. Without standards, the construction environment becomes fragmented, inconsistent, and prone to inefficiencies. With standards, however, designers and constructors share a common understanding, enabling seamless integration between design intent and construction practice. Thus, standardisation theory is not merely relevant to buildability; it is essential for achieving reliable, efficient, and high-quality construction outcomes.

3.7.4 Systems Theory

Systems theory, developed by German biologist Ludwig von Bertalanffy in the 1930s, emerged as a response to reductionist scientific approaches that analysed phenomena by breaking them into isolated parts. Bertalanffy argued instead that many phenomena, biological, social, organisational, and technological, could only be understood as systems: coherent wholes composed of interacting components. A system is therefore defined as an entity with identifiable boundaries separating internal and external elements, allowing the observer to recognise inputs entering the system and outputs emerging from it. Instead of viewing components individually, systems theory focuses on the dynamic relationships and interactions among them. This holistic outlook provides a more complete understanding of how an entity functions, adapts, and produces outcomes.

Systems theory operates at multiple levels of observation. First, it considers the external environment, which includes forces, conditions, and constraints that influence the system. Second, it examines the organisation or project team as a system with roles, processes, and structures. Third, it accounts for human participants, their behaviours, decisions, interactions, and contributions. By integrating these layers, systems theory explains how complex processes such as design coordination, project management, and construction planning unfold within the broader built environment.

Central to systems theory are the components of inputs, throughputs, outputs, and feedback. In the context of buildability assessment, inputs include the demands, resources, constraints, project information, and knowledge that enter the design-management subsystem. Throughputs refer to the processes through which this information is evaluated, coordinated, rationalised, and refined. These may include design reviews, constructability workshops, multidisciplinary coordination meetings, risk assessments, and cost–benefit analyses. Through these processes, designers and constructors clarify production information, resolve conflicts, develop alternative solutions, and structure design decisions that support efficient construction.

Outputs represent the results of these processes: updated drawings, improved detailing, coordinated specifications, construction methodologies, buildability reports, and other deliverables that guide the work of contractors. Effective outputs are those that reduce uncertainties, minimise errors, and enhance construction efficiency.

Feedback plays a critical role throughout the system. Negative feedback helps maintain stability by identifying errors, inconsistencies, or deviations from expected performance and enabling corrective action. Positive feedback, by contrast, encourages improvement and innovation, such as modifying design standards, adopting new technologies, or refining processes based on lessons learned. Feedback loops allow systems to adapt dynamically rather than remain static, contributing to continuous improvement in construction planning and performance. Figure 3.3 illustrates a typical systems theory framework, showing how inputs are transformed through processes and shaped by feedback.

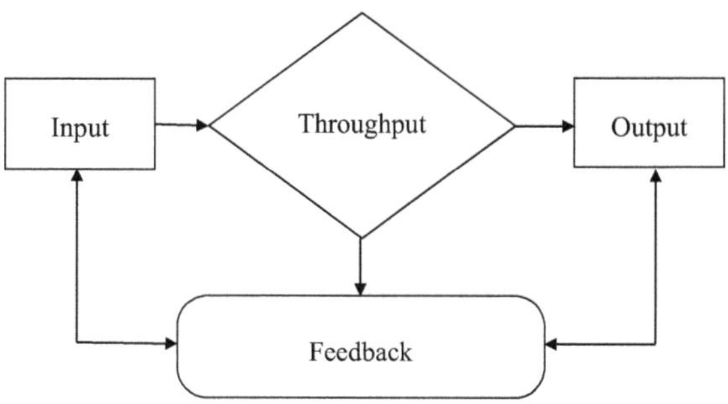

Figure 3.3 System theory framework.

For buildability assessment, maintaining a proper balance among inputs, processes, outputs, and feedback is essential. Without clear inputs, the assessment becomes unfocused. Without structured throughput processes, issues remain unresolved. Without meaningful outputs, construction teams lack actionable guidance. Without feedback, the system fails to adapt or improve.

Systems theory is therefore highly relevant to buildability because buildability itself is a systemic activity. It requires coordinated interactions among designers, builders, consultants, regulators, and clients. It depends on continuous information flow, decision-making, and iterative refinement. Applying systems theory ensures that buildability assessment is treated not as a one-off task, but as an organised, cyclical, and adaptive process aimed at improving overall construction project performance.

3.7.5 Signalling Theory

Signalling Theory, originally developed in economics by Michael Spence (1973), explains how one party, known as the signaller, conveys credible information to another party, the receiver, in situations where information is unevenly distributed. The theory focuses on how signals, whether intentional or unintentional, reduce uncertainty, shape expectations, and influence decision-making. Within the construction context, where multiple stakeholders interact under varying levels of knowledge, signalling plays a critical role in enhancing buildability throughout the project life cycle.

In construction projects, buildability depends heavily on clear, reliable communication of design intentions, constructability requirements, resource capabilities, and potential constraints. However, information asymmetry often exists between designers and contractors, between clients and consultants, or between main contractors and subcontractors. Signalling Theory provides a framework for understanding how these parties use specific mechanisms, such as drawings, specifications, standards, performance records, and qualifications, to communicate competence and reliability. These signals help reduce uncertainty, foster trust, and align expectations, thereby improving buildability outcomes.

For example, designers may provide explicit constructible details, detailed drawings, and rationalised specifications as signals of their understanding of construction processes. Such signals help contractors interpret design intentions more accurately and reduce ambiguity during construction. Similarly, contractors may signal their buildability expertise through demonstrated capability, past project performance, adoption of innovative construction methods, or compliance with industry-recognised standards. These signals influence client decisions during tendering and ensure that the selected contractor possesses the technical competence necessary to deliver a buildable project.

Moreover, signalling supports early collaboration, which is fundamental to achieving buildability. Early contractor involvement, for instance, allows contractors to signal their expertise by providing practical input into design development. Their feedback, relating to construction sequencing, material selection, or modularisation opportunities, serves as a signal of constructability knowledge that helps designers adjust their solutions to enhance ease of construction. In this way, signalling reduces the risk of design-related delays, cost overruns, and rework.

In the context of quality and compliance, signalling also plays an essential role. Adherence to standardised procedures, certification to ISO or industry standards, and use of recognised buildability assessment tools signal that a firm prioritises efficiency and quality. These signals reassure clients and regulators that the construction process will align with established best practices, thereby strengthening buildability. Even the use of Building Information Modelling can be understood through the lens of signalling: providing transparent, coordinated, and clash-free models signals a high level of design integration and constructability awareness.

Importantly, ineffective or misleading signals can negatively impact buildability. Vague drawings, poorly coordinated design information, or exaggerated claims of capability create misinterpretation, leading to inefficiencies, errors, and conflicts. Therefore, signalling must be clear, consistent, and credible to support buildability effectively.

In summary, Signalling Theory provides a valuable lens for understanding the communication dynamics that influence buildability. By reducing information asymmetry, enhancing trust, and improving coordination among stakeholders, effective signalling ensures that design decisions, construction plans, and project delivery processes align with the core objective of achieving efficient, safe, and high-quality construction outcomes.

3.7.6 Agency Theory

Agency theory, developed by economists such as Jensen and Meckling (1976), examines the relationships in which one party, the principal, delegates work to another party, the agent. The theory focuses on the challenges that arise due to information asymmetry, conflicting interests, and differing risk preferences between principals and agents. Within the construction industry, these dynamics are particularly significant because project delivery relies on a complex network of delegated responsibilities among clients, consultants, designers, builders, contractors, and subcontractors. Understanding these relationships through agency theory provides valuable insights into how buildability can be improved or compromised.

In the context of buildability, the principal is often the client (or project owner) who relies on agents such as designers and contractors to deliver a project that is functional, cost-effective, and efficient to construct. However, agents may not always act in the best interest of the principal. Designers, for instance, may prioritise architectural expression or innovation over construction simplicity. Contractors, in turn, may focus on short-term cost savings rather than long-term project efficiency. These misalignments can negatively affect buildability by resulting in overly complex designs, unclear documentation, impractical detailing, or construction methods that do not optimise labour, materials, or time.

Agency theory highlights the importance of mechanisms that align the goals of principals and agents to enhance buildability. One such mechanism is the development of clear contractual frameworks that specify buildability requirements.

For example, performance-based contracts, design-and-build arrangements, or integrated project delivery systems encourage agents to consider construction efficiency during design and execution. These contractual structures reduce information gaps and foster shared responsibility for achieving buildability objectives.

Monitoring and incentive structures are also central to agency theory and have direct relevance to buildability. For instance, regular design reviews, constructability analyses, and buildability assessments serve as monitoring tools that reduce information asymmetry between agents and principals. Incentives, such as performance bonuses for achieving buildable designs, reducing rework, or meeting time and cost targets, motivate agents to act in alignment with the principal's interests. These mechanisms ensure that buildability is not treated as an optional consideration but as a priority embedded in the project's governance framework.

The theory also emphasises the role of transparency in mitigating agency problems. When designers share detailed and coordinated drawings or when contractors provide transparent cost and scheduling information, the likelihood of misunderstandings and conflicting interpretations is reduced. Such transparency supports better decision-making and improves buildability by ensuring that project teams operate based on accurate, timely, and comprehensive information.

Furthermore, agency theory sheds light on the value of early collaboration. When contractors are engaged early in the design process, information asymmetry is minimised. Their input can influence design decisions to improve constructability, mitigate risks, and reduce downstream inefficiencies. This early involvement aligns interests and strengthens mutual accountability.

In summary, agency theory provides a robust framework for understanding how relationships, incentives, and information flows influence buildability. By addressing agency problems through clear contracts, effective monitoring, aligned incentives, and transparent communication, project stakeholders can enhance buildability and ultimately achieve more efficient, cost-effective, and high-quality construction outcomes.

3.7.7 Transaction Cost Economics

TCE, pioneered by Ronald Coase (1937) and later expanded by Oliver Williamson (1975, 1985), explains how economic actors organise transactions in ways that minimise the costs associated with exchanging goods, services, or information. These "transaction costs" include negotiation, contracting, monitoring, coordination, enforcement, and the costs arising from uncertainty or opportunistic behaviour. In construction, an industry characterised by fragmentation, temporary organisations, and high interdependence, TCE provides a powerful lens for understanding how buildability can be enhanced or undermined through the structuring and management of project transactions.

Buildability depends heavily on the quality of interactions among project stakeholders, including clients, designers, contractors, and suppliers. Poor communication, unclear design documentation, disputes, and coordination failures often increase transaction costs, resulting in delays, rework, variations, and reduced efficiency. According to TCE, improving buildability requires reducing these transaction costs by designing project structures, processes, and relationships that support smooth, reliable, and efficient exchanges.

One core concept in TCE is asset specificity, the degree to which certain resources, skills, or knowledge are unique to a particular project. In buildability, high asset specificity is reflected in custom designs, specialised construction techniques, or unique detailing that may be difficult or costly to execute. Highly bespoke designs increase transaction costs because they require intensive negotiation, closer supervision, more contract variations, and greater risk of disputes. By contrast, buildable designs that promote standardisation, modularity, or repeatability reduce asset specificity, thereby lowering contracting and coordination costs.

Uncertainty, another key TCE concept, plays a major role in buildability. Uncertain or ambiguous design information, incomplete drawings, or unrealistic specifications can lead to unexpected complications during construction. This uncertainty increases transaction costs by requiring additional clarifications, design revisions, or changes in work methods. Buildability assessment acts as a mechanism to reduce uncertainty by ensuring that designs are practical, clear, and construction-friendly before work begins. When uncertainty is minimised, project stakeholders spend less time resolving ambiguities and more time executing efficiently.

TCE also emphasises opportunism, the risk that parties may act in self-interest with guile. In construction, opportunistic behaviours may include withholding information, inflating variation prices, or deliberately designing overly complex details to safeguard professional territory. Buildability measures, such as collaborative design reviews or early contractor involvement, reduce opportunities for such behaviour by creating transparency and shared accountability. When the environment promotes trust and cooperation, transaction costs decrease, and buildability improves.

Governance structures, whether market-based contracts, hierarchical controls, or hybrid forms, are also central to TCE. Buildability benefits from governance arrangements that facilitate coordination and knowledge-sharing. For example, design-and-build procurement, alliancing, and integrated project delivery reduce the number of fragmented transactions by bringing key actors together under a shared contractual framework. This alignment minimises the transaction costs arising from conflicting goals or adversarial relationships.

In summary, TCE highlights that buildability is not only a technical or design issue but also an economic and organisational one. By reducing transaction costs related to coordination, uncertainty, asset specificity, and opportunism, construction projects can achieve more efficient, predictable, and cost-effective outcomes. Thus, TCE offers a robust theoretical foundation for understanding and enhancing buildability in modern construction practice.

3.8 Policy and Regulatory Context for Buildability

The policy and regulatory environment plays a fundamental role in shaping how buildability principles are understood, adopted, and implemented within the construction industry. Regulatory frameworks act as the backbone of industry practice, setting the minimum standards that influence design processes, construction methods, procurement strategies, and overall project governance. Through building codes, planning laws, occupational health and safety regulations, and procurement policies, government institutions ensure that construction activities align with national priorities and public interests. These rules and guidelines establish a structured environment in which designers and contractors must operate, thereby encouraging consistent application of buildability principles.

Strong policy support is essential because it ensures that buildability is not treated as an optional consideration but rather as a mandatory component of project development. Policies that promote design simplification, standardisation, safe construction methods, and early stakeholder collaboration encourage project teams to adopt practices that enhance constructability. When regulatory bodies impose clear requirements, such as coordination protocols, safety design obligations, and quality assurance procedures, they provide an enabling environment where buildability becomes embedded in professional practice.

Moreover, well-formulated policies drive innovation by encouraging the use of modern construction techniques, digital technologies, and efficient material systems. Governments that incentivise prefabrication, modular construction, and performance-based design indirectly strengthen buildability outcomes by making these approaches more accessible and economically viable. Policy frameworks also contribute to economic efficiency by reducing waste, minimising design-related errors, and improving project delivery timeframes. At the same time, regulations aligned with environmental goals promote sustainability by encouraging resource-efficient designs, reduced carbon footprints, and eco-friendly construction methods.

In essence, the regulatory environment ensures that buildability principles are systematically integrated into industry operations, thereby improving safety, productivity, and long-term performance in the built environment.

3.8.1 Building Codes and Technical Regulations

Building codes are among the most powerful regulatory mechanisms shaping buildability outcomes in the construction sector. As legally enforceable documents, they specify the minimum performance standards that all building works must satisfy, including requirements related to structural integrity, fire safety, accessibility, energy efficiency, durability, and health protection. Examples include the International Building Code (ICC, 2021), the UK Building Regulations (HM Government, 2010), and comparable national frameworks worldwide. Although these codes are primarily formulated to ensure public safety and welfare, they significantly influence buildability by defining the parameters within which designers, engineers, builders, and contractors must develop solutions.

Clear, consistent, and well-structured building codes enhance buildability by simplifying design decision-making and reducing uncertainties during both design and construction. When requirements are explicit, project teams can more easily develop standardised details, repetitive structural solutions, and modular configurations that align with regulatory expectations. This promotes efficiency, minimises design errors, and reduces on-site rework, key principles of good buildability. Coherent regulatory frameworks contribute to greater predictability in construction processes, facilitating better coordination between design and site implementation.

Moreover, building codes that incorporate performance-based provisions rather than prescriptive rules may further enhance buildability by giving project teams the flexibility to adopt innovative construction technologies, including prefabrication, automation, digital modelling, and advanced materials. Such flexibility encourages design rationalisation and optimisation while still maintaining compliance with essential safety requirements.

On the other hand, inconsistent, ambiguous, or rapidly changing building codes can undermine buildability. Unclear regulations may force designers to make overly conservative assumptions or produce unnecessarily complex details to ensure compliance, both of which impede efficient construction. Frequent regulatory changes also lead to additional design revisions, coordination delays, and higher compliance costs, all of which negatively affect buildability. Therefore, the effectiveness of building codes as a buildability enhancer depends largely on their clarity, stability, and alignment with contemporary construction practices and technologies.

3.8.2 Planning Regulations and Land-Use Policies

Buildability is also strongly influenced by statutory planning controls, zoning regulations, environmental approvals, and development guidelines, all of which shape the physical, spatial, and environmental parameters within which a project must be conceived and executed. These frameworks regulate essential aspects of development such as allowable building heights, plot ratios, densities, setbacks, land-use designations, site access, and obligations for environmental protection, heritage conservation, or community amenity. As a result, planning policy represents a foundational layer that either enhances or constrains buildability.

In greenfield developments, where land is largely undeveloped, planning frameworks tend to be more straightforward and flexible. Such sites often fall within newly zoned or master-planned areas where regulatory requirements are clearly articulated and long-term development intentions have already been established. De Sousa (2000) argues that this clarity facilitates smoother integration of design and construction processes, providing opportunities for efficient site layout, modular construction approaches, rationalised service routes, and optimised construction logistics. The relative absence of pre-existing structures, contamination risks, or heritage constraints further enables designers and contractors to apply buildability principles with fewer restrictions.

In contrast, brownfield developments must navigate significantly more complex planning and regulatory environments. These sites may be constrained by existing land uses, protected heritage structures, community sensitivities, environmental remediation requirements, or intricate land-title and ownership conditions. Consequently, developers and design teams are often required to undergo additional layers of regulatory review, including heritage permissions, environmental impact assessments, utility coordination approvals, or traffic impact analyses. These processes can restrict design flexibility, extend project timelines, add compliance costs, and necessitate more adaptive construction strategies, factors that can challenge buildability if not effectively managed.

Therefore, predictable, transparent, and well-coordinated planning policies are essential in supporting buildability. Efficient approval processes, consistent zoning interpretations, and integrated environmental and heritage guidelines help reduce regulatory uncertainty, allowing project teams to develop designs that are constructible, economical, and aligned with statutory requirements.

3.8.3 Occupational Health and Safety Regulations

Health and safety regulations play a critical role in reinforcing buildability by requiring construction processes to be planned, designed, and executed in ways that minimise risks and promote safe working environments. These regulations function not only as compliance obligations but also as mechanisms that push the industry toward more systematic, predictable, and construction-friendly design solutions. A central example is the UK's Construction (Design and Management) Regulations 2015 (HSE, 2015), which explicitly impose duties on designers to eliminate, reduce, or control foreseeable hazards during the earliest stages of project conception. By compelling designers to consider how their decisions influence construction safety, the regulations effectively embed buildability considerations into the design process.

This legislative emphasis aligns closely with established research showing strong links between safe design and improved buildability. Heong and Abas (2021) demonstrated that simplified design details, reduced on-site labour intensity, and the adoption of repeatable construction processes can significantly reduce accident risks. Strategies that promote safety, such as design standardisation, rationalised building layouts, reduced dependence on complex temporary works, and increased use of prefabricated or modular components, also support buildability by making construction more efficient, predictable, and less prone to errors.

As global safety regulations become more stringent, the interdependence between safety and buildability grows stronger. Designers must increasingly demonstrate that their design choices facilitate safe construction sequencing, secure access for workers, manageable lifting operations, and logical build processes. Contractors, in turn, rely on buildable designs to minimise the risk of safety incidents, delays, and regulatory non-compliance. Therefore, health and safety legislation acts as a powerful driver for integrating buildability principles, encouraging all project participants to adopt construction-friendly design practices that enhance both performance and protection.

3.8.4 Procurement Policies and Contractual Frameworks

Public-sector procurement guidelines are powerful instruments that shape the extent to which buildability is embedded in project delivery. Increasingly, governments recognise that traditional procurement models, characterised by rigid separation between design and construction, often hinder the integration of practical construction knowledge during early project phases. As a result, many jurisdictions now promote or mandate procurement approaches that prioritise collaboration, early engagement, and shared responsibility. Notable among these are early contractor involvement, integrated project delivery, and design-and-build arrangements, all of which enable contractors to contribute constructability expertise before design decisions are finalised. Such early input reduces downstream rework, prevents design–construction conflicts, enhances risk forecasting, and ultimately improves project outcomes.

Collaborative procurement models supported by public-sector guidelines also help address chronic fragmentation within the construction industry. By encouraging integrated teams, linking clients, designers, engineers, builders, contractors, and specialist subcontractors, public procurement policies facilitate knowledge-sharing and coordinated decision-making. This integration directly strengthens buildability by ensuring that design choices reflect real construction constraints, resource availability, site conditions, and safety considerations. Furthermore, collaborative contracting reduces adversarial relationships and lowers transaction costs associated with disputes, claims, and contractual ambiguities.

Policy frameworks increasingly incorporate performance-based mechanisms that reward innovation, efficiency, sustainability, and reliable project delivery. Incentive clauses, key performance indicators, and risk-sharing models encourage project teams to pursue design simplification, standardisation, modularisation, and other strategies that enhance buildability. Government agencies also establish procurement guidelines that evaluate contractors not only on cost but also on technical expertise, safety history, and proven ability to deliver buildable designs. Such measures elevate buildability from a discretionary consideration to a formal evaluation criterion within bidding and selection processes.

Overall, public-sector procurement policies act as catalysts for buildability improvement by fostering collaboration, rewarding innovation, and ensuring that construction knowledge is fully integrated into project planning and design.

3.8.5 Sustainability and Environmental Regulations

Sustainability policies increasingly influence buildability by reshaping how buildings are designed, constructed, and managed across their life cycles. Contemporary frameworks, such as BREEAM in the UK, LEED in the United States, Green Star in Australia, and various national green building codes, emphasise energy efficiency, reduced material waste, responsible resource use, and lower carbon emissions. These requirements encourage design teams and contractors to adopt construction methods that inherently support buildability, such as prefabrication, modular construction, and lean construction practices. Prefabrication, for example, reduces on-site labour intensity, minimises waste, shortens construction timelines, and improves quality control, outcomes that align closely with buildability objectives.

Sustainability regulations also promote material optimisation and responsible sourcing, which often necessitate early coordination between designers, suppliers, and contractors. Such early decision-making is a core buildability principle, as it prevents later-stage design revisions and reduces construction uncertainties. Furthermore, stringent environmental requirements, such as mandatory waste management plans, stormwater controls, carbon reporting, and site disturbance minimisation, demand careful planning of construction logistics. This emphasis on planning enhances buildability by reinforcing the need for clear sequencing, efficient site layout, and integration of construction expertise during the design stage.

Energy-efficiency mandates likewise influence buildability by encouraging simplified building forms, standardised components, and high-performance materials that reduce thermal bridging and improve insulation. These features often require precise detailing and coordinated design information, which in turn supports ease of construction. Sustainable design guidelines also promote durable, low-maintenance materials, which help to reduce errors and rework during construction.

In summary, sustainability policies actively shape buildability by driving early-stage integration, encouraging streamlined construction processes, and promoting the adoption of innovative and resource-efficient techniques. As environmental regulations become more rigorous, the synergy between sustainability and buildability will continue to strengthen, making both essential components of modern construction practice.

3.8.6 Government-Led Buildability Initiatives

Some jurisdictions have introduced explicit policies and regulatory tools designed to measure, monitor, and enforce buildability in construction projects. A leading example is Singapore's Buildable Design Appraisal System (BDAS), introduced by the Building and Construction Authority. Under BDAS, all major building projects are required to achieve a minimum buildability score, with higher scores awarded for features such as simplified structural systems, modular layouts, standardised components, and extensive use of prefabrication technologies (BCA, 2020). This policy has transformed the Singapore construction industry by incentivising designers to prioritise constructability from the earliest design stages, and by promoting a culture of standardisation, off-site manufacturing, and integrated design thinking.

The success of BDAS demonstrates how regulatory frameworks can drive measurable improvements in productivity, safety, quality, and sustainability. Buildability assessments encourage design teams to minimise complexity, reduce labour-intensive processes, and adopt technologies such as Prefabricated Prefinished Volumetric Construction and precast systems, methods strongly aligned with global trends toward industrialised construction. As a result, Singapore has become an international benchmark for buildability-focused regulation.

Other countries have introduced similar, though less formalised, initiatives. For example, the UK and Australia promote buildability through government guidance on modern methods of construction, early contractor involvement, and standardised component libraries. In Scandinavia, policies supporting industrialised building systems indirectly improve buildability by encouraging modularisation and advanced digital coordination. These approaches reflect an increasing global recognition that buildability is not only a technical design issue but also a strategic policy mechanism for improving construction-sector performance.

Overall, the adoption of targeted buildability policies underscores its rising importance in national agendas focused on productivity, safety, innovation, and environmental stewardship. As more jurisdictions confront labour shortages, rising costs, and sustainability pressures, formal buildability frameworks are likely to become increasingly widespread.

3.8.7 Implications for Construction Practice

The regulatory environment serves as a foundational force shaping how buildability is interpreted, prioritised, and implemented across the construction industry. Coherent, well-aligned policies create an enabling environment in which designers, builders, contractors, and clients are encouraged, or required, to embed buildability principles from project inception. When regulations clearly articulate expectations for design efficiency, safety integration, procurement coordination, and sustainability performance, they help standardise industry practices and reduce ambiguity in decision-making. This alignment supports smoother project delivery, reduces rework, enhances productivity, and promotes a culture of proactive planning and interdisciplinary collaboration.

In contrast, fragmented, outdated, or inconsistent regulations can hinder buildability efforts by creating uncertainty, conflicting requirements, or administrative burdens that disrupt early design coordination. Regulatory gaps may lead to inconsistent interpretations of buildability requirements, while overlapping jurisdictions or slow approvals can delay the design process and undermine opportunities for early contractor involvement. Such weaknesses often result in suboptimal construction solutions, increased costs, and reduced innovation as stakeholders focus on compliance rather than optimisation.

Global policy trends indicate that buildability will continue to gain prominence as governments address pressing industry challenges. Productivity stagnation, labour shortages, climate-change obligations, and the shift toward digital construction technologies, such as Building Information Modelling, digital twins, and off-site manufacturing, are pushing regulators to integrate buildability more explicitly into statutory requirements and industry guidelines. Emerging regulatory frameworks increasingly emphasise design-for-manufacture-and-assembly, standardisation, modular construction, and digital workflows, all of which strengthen buildability outcomes.

As construction becomes more technologically advanced and sustainability-driven, regulatory systems are expected to evolve toward greater integration, transparency, and performance-based requirements. Ultimately, strong and forward-looking regulatory support will be essential in embedding buildability as a routine component of design and project governance, ensuring that the built environment is delivered more efficiently, safely, and sustainably.

3.9 Chapter Conclusion

This chapter has laid the conceptual and theoretical foundation for understanding buildability as both a practical construction principle and a multidisciplinary theoretical construct. By examining definitions, principles, attributes, and influencing factors, this chapter has clarified that buildability extends beyond ease of construction; it encompasses the systematic alignment of design, procurement, construction processes, regulatory compliance, and lifecycle considerations to achieve optimal project performance.

The exploration of definitions and principles revealed that buildability is inherently concerned with simplifying construction processes, enhancing coordination, minimising uncertainty, and promoting efficient resource utilisation. Across project phases, from conceptual design to execution and operation, buildability attributes evolve, yet remain interconnected. Early-stage decisions have disproportionate influence on downstream outcomes, reinforcing the importance of integrating constructability considerations at the earliest possible stage. Where such integration is absent, buildability problems emerge, often manifesting as design conflicts, delays, rework, cost overruns, safety risks, and disputes.

By analysing the factors affecting buildability and the causes and impacts of related problems, this chapter demonstrated that buildability challenges are rarely technical alone. Organisational fragmentation, inadequate communication, limited contractor input, regulatory complexity, and misaligned incentives frequently undermine constructability. These issues highlight the systemic nature of buildability and the need for coordinated strategies to address them. The strategies outlined, such as early stakeholder engagement, design simplification, standardisation, improved documentation, and collaborative procurement, serve as practical pathways to enhance project performance.

The distinction between greenfield and brownfield contexts further emphasised that buildability is context-sensitive. Greenfield projects offer greater flexibility in planning and sequencing, whereas brownfield developments must contend with constraints such as existing infrastructure, operational continuity, environmental risks, and regulatory complexity. Recognising these contextual differences is essential for tailoring buildability strategies to specific project environments.

The theoretical perspectives discussed in this chapter deepen the understanding of buildability beyond practice-based observations. Rationalisation, simplicity, and standardisation theories explain how efficiency and repetition reduce complexity and variability. Systems theory highlights the interdependence of project components and stakeholders, reinforcing the need for integrated decision-making. Signalling theory, agency theory, and TCE illuminate the behavioural and contractual dynamics that influence collaboration, information sharing, and risk allocation. Collectively, these theoretical lenses position buildability as a socio-technical construct shaped by incentives, governance structures, and organisational relationships as much as by engineering considerations.

Finally, the policy and regulatory context underscores the broader institutional environment within which buildability operates. Building codes, planning regulations, occupational health and safety requirements, procurement policies, and sustainability frameworks all shape design and construction decisions. Government-led initiatives and regulatory reforms can either enable or constrain buildability, depending on how effectively they align compliance objectives with practical constructability outcomes.

In conclusion, buildability is both a conceptual framework and an applied strategy grounded in theory, practice, and policy. Understanding its principles, contextual variations, systemic influences, and regulatory dimensions provides a comprehensive foundation for improving construction performance. This chapter establishes that enhancing buildability requires not only technical optimisation but also organisational alignment, theoretical insight, and supportive institutional frameworks, elements that collectively shape the efficiency, resilience, and sustainability of construction projects.

Bibliography

Adams, S. (1989). *Practical buildability*. Butterworths, p. 2.

Afolabi, A., & Oyeyipo, O. (2017). The perception of future decision makers on the building profession. *Malaysian Construction Research Journal*, 21(1), 55–73.

Aina, O.O. (2015). Buildability problems in construction projects in Nigeria: Causes and impacts. *European Scientific Journal*, 11(27), 353–366.

Aina, O.O., & Wahab, A.B. (2011). An assessment of buildability problems in the Nigerian construction industry. *Global Journal of Research Engineering*, 11(2), 42–52.

Al-Faryan, M.A.S. (2024). Agency theory, corporate governance and corruption: An integrative literature review approach. *Cogent Social Sciences*, 10(1). https://doi.org/10.1080/23311886.2024.2337893

Althoey, F., Houda, M., Waqar, A., & Nisar, S. (2024). Comprehensive framework for integrated construction platforms (ICP) implementation and its effect on project success. *Ain Shams Engineering Journal*, 15(12), 103152. https://doi.org/10.1016/j.asej.2024.103152

Alzayd, R. (2016). A constructability methodology to integrate planning and design for workers' safety in construction management. PhD. Thesis, University of Florence and Technische Universität Braunschweig.

Arditi, D., & Mochtar, K. (2000). Trends in productivity improvement in the US construction industry. *Construction Management and Economics*, 18(1), 15–27. https://doi.org/10.1080/014461900370915

Bafera, J., & Kleinert, S. (2023). Signaling theory in entrepreneurship research: A systematic review and research agenda. *Entrepreneurship Theory and Practice*, 47(6), 2419–2464. https://doi.org/10.1177/10422587221138489

Bamisile, A. (2004). *Building production management*. Foresight Press Ltd.

BCA. (2020). *Code of practice on buildability*. Building and Construction Authority, Singapore. Available at: https://isomer-user-content.by.gov.sg/338/aa52817a-ac63-4f2a-8a06-ece42005f0ed/cop-on-buildability-2020e7f7044a2d8e4a2ca7daacd64c3f0691.pdf (accessed May 13, 2026).

Bello, W.A. (2017). Project performance diagnostics: A model for assessing construction project performance in Nigeria. *A thesis submitted in partial fulfilment of the requirements for the Degree of Doctor of Philosophy*. School of the Built Environment, University of Salford, Manchester. Available at: https://usir.salford.ac.uk/id/eprint/44793/1/Bello%20PhD%20%20PPDM%20Final%20Submission%202017.pdf

Bendickson, J., Muldoon, J., Liguori, E.W., & Davis, P.E. (2016). Agency theory: Background and epistemology. *Journal of Management History*, 22(4), 437–449. https://doi.org/10.5465/ambpp.2016.12665abstract

Cheetham, D.W., & Lewis, J. (2001). Productivity, buildability and constructability: Is work study the missing link? In: Akintoye, A. (Ed.) *17th annual ARCOM conference*, 5–7 September 2001, University of Salford. Association of Researchers in Construction Management, 1, 271–280.

Chen, S.E., McGeorge, W.D., & Ostwald, M.J. (1993). The role of information management in the development of a strategic model of buildability. In: *Management of information technology for construction conference*, Singapore.

Cherepanov, V., Feddersen, T., & Sandroni, A. (2013). Rationalization. Theoretical Economics, 8: 775-800. https://doi.org/10.3982/TE970

Chiang, Y.H., Tao, L., & Wong, F.K.W. (2015). Causal relationship between construction activities, employment and GDP: The case of Hong Kong. *Habitat International*, 46, 1–12. https://doi.org/10.1016/j.habitatint.2014.10.016

Choudhury, M.M. (2024). Signaling theory: An approach to organizational behavior research. *Journal of Accounting Business and Management (JABM)*, 31(2), 98–120. https://doi.org/10.31966/jabminternational.v31i2.1199

Coase, R.H. (1937). The nature of the firm. *Economica*, 4, 386–405.

Construction Industry Institute (CII). (1986). *Buildability: A primer*, 3(1). University of Texas at Austin, TX.

Construction Industry Institute (CII). (1993). *Preview of constructability implementation.* CII, Austin, TX.

Construction Industry Research and Information Association (CIRIA). (1983). *Buildability: An assessment.* CIRIA Publications, Special Report, London, p. 26.

Cox, L., & Rodway-Dyer, S. (2023). The underappreciated value of brownfield sites: motivations and challenges associated with maintaining biodiversity. *Journal of Environmental Planning and Management*, 66(9), 2009–2027. https://doi.org/10.1080/09640568.2022.2050683

Crowther, P. (2002). *Design for buildability and the deconstruction consequences*. In A. Chini & F. Schultmann (Eds.), *Design for Deconstruction and Materials Reuse* (CIB Publication No. 272, Proceedings of the CIB Task Group 39 – Deconstruction Meeting, Karlsruhe, Germany, 9 April 2002). International Council for Research and Innovation in Building and Construction (CIB).

Dang, T.H.G., & Low, S.P. (2011). Role of construction in economic development: Review of key concepts in the past 40 years. *Habitat International*, 35(1), 118–125. https://doi.org/10.1016/j.habitatint.2010.06.003

Dantata, S.A. (2007). General overview of the Nigerian construction industry. Project Submitted to the Department of Civil & Environmental Engineering, in Partial Fulfillment of the Requirements for the Degree of Master of Engineering in Civil and Environmental Engineering at the Massachusetts Institute of Technology.

De Sousa, C. (2000). Brownfield redevelopment versus greenfield development: A private sector perspective on the costs and risks associated with brownfield redevelopment in the Greater Toronto Area. *Journal of Environmental Planning and Management*, 43(6), 831–853. https://doi.org/10.1080/09640560020001719

Dessalles, J.-L. (2011). Algorithmic simplicity and relevance. In: Dowe, D.L. (Ed.) *Proceedings of the solomonoff 85th memorial conference*. https://doi.org/10.1007/978-3-642-44958-1_10

Dorcas, O.A.B., Elkanah, A.O., & John, O.O. (2019). Most critical factors responsible for poor project quality performance in building construction industry (A case study of three major cities in Nigeria). *European International Journal of Science and Technology*, 8(2), 1–14.

Eisenhardt, K.M. (1989). Agency theory: An assessment and review. *The Academy of Management Review*, 14(1), 57–74. https://doi.org/10.2307/258191

Emmerson, H. (1962). *Survey of problems before the construction industries. A report prepared for the Ministry of Works*. HMSO, London.

Ferguson, I. (1989). *Buildability in practice*. Mitchell's Professional Library.

Fisher, D.J., & Rajan, N. (1996). Automated buildability: Analyze of work-zone traffic control planning. *Journal of Construction, Engineering and Management*, 122(1), 36–43.

Foliente, G.C. (2000). Developments in performance-based building codes and standards. *Forest Products Journal*, 50(7), 12–21.

Fomin, V., Keil, T., & Lyytinen, K. (2003). Theorizing about standardization: Integrating fragments of process theory in light of telecommunication standardization wars. *Sprouts: Working Papers on Information Environments, Systems and Organizations*, 3(1), 29–60. Available at: http://sprouts.aisnet.org/3-10 (accessed July 3, 2021).

Griffith, A. (1987). *An investigation into factors influencing buildability and levels of productivity for application to selecting alternative design solutions – A preliminary report*. In *Proceedings of the 5th CIB W65 Symposium on Organisation and Management of Construction* (Vol. 2, pp. 646–657). Chartered Institute of Building.

Hammond, E.B., Coulon, F., Hallett, S.H., Thomas, R., Hardy, D., & Beriro, D.J. (2023). Digital tools for brownfield redevelopment: Stakeholder perspectives and opportunities. *Journal of Environmental Management*, 325, Part A, 116393. https://doi.org/10.1016/j.jenvman.2022.116393

Health and Safety Executive (HSE). (2015). *Managing health and safety in construction: Construction (design and management) regulations 2015*. Crown. ISBN 9780717666263. Available at: https://www.hse.gov.uk/pubns/priced/l153.pdf (accessed December 9, 2025).

Heong, Y.W., & Abas, N.H. (2021). Preventive measures of struck-by accidents at construction site: Perspectives from construction personnel in Johor. In: *Conference: 2021 2nd international hybrid conference on durability of building and infrastructures for smart city*, (DuraBI) October 2021, 1375–1379.

HM Government. (2010). *The building regulations*. UK Government. Available at: https://assets.publishing.service.gov.uk/media/5a80437640f0b623026927b2/BR_PDF_AD_A_2013.pdf (accessed December 9, 2025).

ICC. (2021). *International building code*. International Code Council. Available at: https://core-docs.s3.amazonaws.com/documents/asset/uploaded_file/3836/Lennox/3042180/2021_International_Building_Code.pdf (accessed December 9, 2025).

Illingworth, J.R. (1984). *Buildability – Tomorrow's need, building technology and management*, pp.16–19.

Isa, R.B., Jimoh, R.A., & Achuenu, E. (2013). An overview of the contribution of construction sector to sustainable development in Nigeria. *Net Journal of Business Management*, 1(1), 1–6.

Jarkas, A.M. (2010). The effects of buildability factors on formwork labour productivity of grade beams. *Revista Ingenieria de Construction*, 25(2), 241–266.

Jensen, M.C., & Meckling, W.H. (1976). Theory of the firm: Managerial behavior, agency costs and ownership structure. *Journal of Financial Economics*, 3(4), 305–360. https://doi.org/10.1016/0304-405X(76)90026-X.

Jones, E. (1908). Rationalisation in everyday life. *Journal of Abnormal Psychology*, 3(3), 161–169.

Kamali, M., Hewage, K., Rana, A., Alam, S., & Sadiq, R. (2025). Advancing urban resilience with modular construction: An integrated sustainability assessment framework. *Resilient Cities and Structures*, 4(2), 46–68. https://doi.org/10.1016/j.rcns.2025.02.006

Koll-Schretzenmayr, M. (1999). From greenfield development to brownfield edevelopment: New challenges for planning process, planning strategies and planning law. *disP - The Planning Review*, 35(139), 43–48. https://doi.org/10.1080/02513625.1999.10556724

Lai, C.H., & Lin, S.H. (2017). *Systems theory*. JohnWiley & Sons, Inc. https://doi.org/10.1002/9781118955567.wbieoc203

Lam, P.T.I., & Wong, F.W.H. (2008). Implementing a buildability assessment model for buildability improvement. *Architectural Science Review*, 51(2), 173–184. https://doi.org/10.3763/asre.2008.5121

Lam, P.T.I., Wong, F.W.H., & Chan, A.P.C. (2006). Contributions of designers to improving buildability and constructability. *Design Studies*, 27(4), 457–479.

Linder, S., & Foss, N.J. (2015). Agency theory. In: Wright, J.D. (Ed.) *International encyclopedia of the social & behavioral sciences* (2nd ed., pp. 344–350). Elsevier. https://doi.org/10.1016/B978-0-08-097086-8.73038-8

Mannontaratana, P., Laovisutthichai, V., Viryasiri, T., Jamnothai, M., & Poonpermpun, W. (2025). Prescriptive-based building regulations: Challenges and recommendations for modernization in the case of Bangkok. *Nakhara: Journal of Environmental Design and Planning*, 24(3), 519. https://doi.org/10.54028/NJ202524519

Matara, M.O.F. (2008). A survey of sources of finance for building construction firms in Kenya. Unpublished master's thesis. University of Nairobi, pp. 1–54.

McKinsey Global Institute. (2017). *Reinventing construction: A route to higher productivity*. McKinsey & Company.

Mbamali, I., Aiyetan, O.A., & Kehinde, J.O. (2005). Building design for buildability: An investigation of the current practice in Nigeria. *Building and Environment*, 40(9), 1267–1274. https://doi.org/10.1016/j.buildenv.2004.10.006

Mele, C., Pels, J., & Polese, F. (2010). A brief review of systems theories and their managerial applications. *Service Science*, 2(1/2), 126–135. https://doi.org/10.1287/serv.2.1_2.126

Mĺkva, M., Prajová, V., Yakimovich, B., Korshunov, A., & Tyurin, I. (2016). Standardization - One of the tools of continuous improvement. In: *International conference on manufacturing engineering and materials, ICMEM 2016*, 6–10 June 2016, Nový Smokovec, Slovakia. *Procedia Engineering*, 149 (2016), 329–332. https://doi.org/10.1016/j.proeng.2016.06.674

Moore, D.R. (1993). Buildability and skill concept packages: Development of a possible design tool: Basic concept of skill modelling is that core movements, common to physical tasks, can be identified, assessed, and placed in a knowledge base. *Building Research & Information*, 21(2), 117–121. https://doi.org/10.1080/09613219308727271

Mydin, S.H., Zin, R.M., Majid, M.Z.A., Zahidi, M., & Memon, A.H. (2011). Buildability attributes at design phase in Malaysian building construction. *International Journal of Sustainable Construction Engineering & Technology*, 2(1), 24–43.

Odekunle, J.F., Akindele, D.B., & Adebayo, G.O. (2019). The problems and challenges of development control in Abeokuta-west zonal planning area, Ogun state, Nigeria. *African Journal of Economics and Sustainable Development*, 2(1), 9–27.

Okeyinka, O.M., Khan, R., Pathirage, C., Mahammedi, C.E.D., & West, A.A. (2023). Critical review of developers' decision criteria for brownfield regeneration: Development of the BRIC index. *Sustainability*, 15, 7105. https://doi.org/10.3390/su15097105

Oladinrin, T.O., Ogunsemi, D.R., & Aje, I.O. (2012). Role of construction sector in economic growth: Empirical evidence from Nigeria. *FUTY Journal of the Environment*, 7(1), 50–60. https://doi.org/10.4314/fje.v7i1.4

Osuizugbo, I.C. (2020a). Improving the performance of building construction firms through addressing the gap of building production management: A new production model approach. *Journal of Engineering, Project, and Production Management*, 10(1), 50–63. https://doi.org/10.2478/jeppm-2020-0007

Osuizugbo, I.C. (2020b). The need for and benefit of buildability analysis: Nigeria as a case study. *Journal of Engineering, Design and Technology*, 19(5), 1207–1230. https://doi.org/10.1108/JEDT-08-2020-0338

Osuizugbo, I.C. (2025a). Buildability-led procurement: A new model for future construction projects delivery. *Construction Innovation: Information, Process, Management*. https://doi.org/10.1108/CI-10-2023-0272

Osuizugbo, I.C. (2025b). Measures for improving the buildability of building designs in construction industry. *Construction Innovation: Information Process Management*, 25(2), 490–509. https://doi.org/10.1108/CI-08-2022-0223

Osuizugbo, I.C., & Ojelabi, A.R. (2020). Building productionmanagement practice in the construction industry in Nigeria. *Engineering Management in Production and Services*, 12(2), 56–73. https://doi.org/10.2478/emj-2020-0011

Pavolová, H., Bakalár, T., Tokarčík, A., Kozáková, Ľ., & Pastyrčák, T. (2021). An economic analysis of brownfield and greenfield industrial parks investment projects: A case study of Eastern Slovakia. *International Journal of Environmental Research and Public Health*, 18(7), 3472. https://doi.org/10.3390/ijerph18073472

Rashid, S.A.U., & Hadikusumo, B.H.W. (2025). Implementation of construction safety laws in public and private projects: Empirical evidence from Bangladesh. *SSRG International Journal of Civil Engineering*, 12(5), 127–141. https://doi.org/10.14445/23488352/IJCE-V12I5P111

Rindfleisch, A. (2020). Transaction cost theory: Past, present and future. *AMS Review*, 10, 85–97. https://doi.org/10.1007/s13162-019-00151-x

Roy, S. (2018). Rationalization. Unit IV, e-content for PMIR-Semester-II_Paper Code-CC-9, Lecture Note, Department of PMIR, Patna University, Patna.

Singhaputtangkul, N., & Low, S.P. (2015). Modelling a decision support tool for buildable and sustainable building envelope designs. *Buildings*, 5, 521–535. https://doi.org/10.3390/buildings5020521

Singhaputtangkul, N., Low, S.P., Teo, A.L., & Hwang, B.G. (2013). Analysis of criteria for decision making to achieve sustainability and buildability in building envelope design. *Architectural Science Review*. https://doi.org/10.1080/00038628.2013.783458

Soewin, E., & Chinda, T. (2018). Factors affecting construction performance: Exploratory factor analysis. *IOP Conference Series: Earth and Environmental Science,* 140, 012102. https://doi.org/10.1088/1755-1315/140/1/012102

Spence, M. (1973). Job market signaling. *The Quarterly Journal of Economics*, 87(3), 355–374. https://doi.org/10.2307/1882010

Tagesse, B. (2017). Assessment of construction performance challenges in selected university building construction projects. Unpublished master's thesis. Addis Ababa University. Available at: http://etd.aau.edu.et/handle/12345678/5437

Trier, M., & Richter, A. (2013). *"I Can Simply…"? Theorizing simplicity as a design principle and usage factor*. In *Proceedings of the 21st European Conference on Information Systems (ECIS 2013)*, Utrecht, The Netherlands, 5–8 June 2013 (Paper 79). Association for Information Systems.

Trigunarsyah, B. (2004). Project owners' role in improving constructability of construction projects: An example analysis for Indonesia. *Construction Management and Economics*, 22, 861–876. https://doi.org/10.1080/01446190410001678107

Vega, D.C., & Keenan, R.J. (2014). Transaction cost theory of the firm and community forestry enterprises. *Forest Policy and Economics*, 42, 1–7. https://doi.org/10.1016/j.forpol.2014.01.006

Williamson, O.E. (1975). *Markets and hierarchies: Analysis and antitrust implications*. The Free Press.

Williamson, O.E. (1979). Transaction-cost economics: The governance of contractual relations. *The Journal of Law & Economics*, 22(2), 233–261. https://www.jstor.org/stable/725118

Williamson, O.E. (1985). *The economic institutions of capitalism*. The Free Press.

Williamson, O.E. (1989). Chapter 3: Transaction cost economics. In: *Handbook of industrial organization* (Vol. 1, pp. 135–182). Elsevier. https://doi.org/10.1016/S1573-448X(89)01006-X

Williamson, O.E. (1998). Transaction cost economics: How it works; where it is headed. *De Economist*, 146, 23–58. https://doi.org/10.1023/A:1003263908567

Williamson, O.E. (2010). Transaction cost economics: The natural progression. *The American Economic Review*, 100(3), 673–690. https://www.jstor.org/stable/27871227

Wong, F., Lam, P., & Shen, L. (2004). A dynamic design management system for improving buildability of construction. In: Khosrowshahi, F. (Ed.) *20th annual ARCOM conference,* Vol. 1, pp. 185–194. Heriot Watt University: Association of Researchers in Construction Management.

Wong, F.W.H., Lam, P.T.I., Chan, E.H.W., & Wong, F.K.W. (2006a). Factors affecting buildability of building designs. *Canadian Journal of Civil Engineering*, 33,795–806. https://doi.org/10.1139/L06-022

Wong, F.W.H., Lam, P.T.I., Chan, A.P.C., & Chan, E.H.W. (2006b). A review of buildability performance in Hong Kong and strategies for improvement. *Surveying and Built Environment,* 17(2), 37–48.

Wong, F.W.H., De Saram, D.D., Lam, P.T.I., & Chan, D.W.M. (2006c). A compendium of buildability issues from the viewpoints of construction practitioners. *Architectural Science Review,* 49(1), 81–90. https://doi.org/10.3763/asre.2006.4910

Wong, F.W.H., Lam, P.T.I., Chan, E.H.W., & Shen, L.Y. (2007). A study of measures to improve constructability. *International Journal of Quality & Reliability Management*, 24(6), 586–601. https://dx. doi.org/10.1108/02656710710757781

Xie, Z., Hall, J., McCarthy, I.P., Skitmore, M., & Shen, L. (2016). Standardization efforts: The relationship between knowledge dimensions, search processes and innovation outcomes. *Technovation. Innovation and Standardization.* https://doi.org/10.1016/j.technovation.2015.12.002

Zhu, Z.M. (2024). The application of transaction cost theory in supply chain management. *Open Journal of Applied Sciences*, 14, 3216–3225. https://doi.org/10.4236/ojapps.2024.1411212

Part II

Applications in Construction Practice

4 Buildability in Construction

4.1 Buildability-Led Procurement

Buildability-Led Procurement (BLP) has emerged as a progressive, relationship-based procurement strategy designed to embed buildability considerations at the core of project planning, design, and delivery. Unlike traditional procurement methods that often separate design and construction functions, BLP seeks to unify the knowledge base of all key stakeholders, designers, builders, contractors, suppliers, and facility managers, through early, continuous, and structured collaboration. The philosophy behind BLP is grounded in principles of cooperation, transparency, early contractor involvement (ECI), and integrated team operation. It represents a strategic shift in how construction projects are conceptualised, evaluated, and executed, placing buildability as a central lens for decision-making across the entire project life cycle.

4.1.1 Foundations and Rationale of BLP

The foundation of BLP lies in the recognition that construction knowledge is most valuable when applied early and consistently. Historically, the construction industry has been challenged by fragmented practices, where designers make decisions without full insight into construction methods, materials, or site-specific constraints. This fragmentation often leads to constructability issues, design inefficiencies, cost overruns, delays, and rework during construction. BLP attempts to remedy these issues by structuring procurement around buildability analysis and collaborative problem-solving. The concept of BLP was first developed by Innocent Chigozie Osuizugbo in 2025. His pioneering work provided the initial theoretical foundation and formal definition of BLP. This contribution marked a major shift in procurement thinking by positioning buildability as the central driver of project decision-making.

The development of BLP as a procurement method is partly attributed to the increasing complexity of construction projects and the diverse stakeholder requirements shaping the built environment. Large and complex projects, such as mega-infrastructure, high-rise commercial structures, and technologically advanced facilities, demand a higher level of integration between design and construction knowledge. As van der Walt et al. (2019) observed, the scale and

DOI: 10.1201/9781003747284-6

complexity of modern projects necessitated the involvement of construction experts early in the project lifecycle. Likewise, Hastie et al. (2017) emphasised that shifting client expectations, constantly evolving regulatory frameworks, and the drive for sustainable and high-performance buildings have triggered the need for procurement approaches that can align diverse objectives while maintaining efficiency and value.

In response to these industry dynamics, BLP provides a structured way to incorporate professional construction knowledge into design decisions, ensuring that project strategies remain grounded in practical, efficient, and feasible construction methodologies. Through this approach, BLP supports the selection of the most beneficial design and construction solutions during conceptual planning and continues to refine these solutions throughout the project lifecycle.

4.1.2 Core Components of BLP

BLP is distinguished by several interrelated components, each contributing to the effective integration of buildability into procurement processes. These components include:

- Buildability – the central guiding principle, ensuring design decisions reflect efficient construction processes.
- Maintainability – ensuring long-term functionality and ease of maintenance is considered during procurement.
- Collaboration – fostering teamwork and transparency across project stakeholders.
- Supply Chain Management – integrating suppliers and specialist subcontractors into early design discussions.
- Risk and Value Management – using structured techniques to optimise project outcomes.

Together, these components create an environment in which procurement decisions are evaluated holistically, balancing technical, economic, environmental, social, and institutional project goals. BLP views buildability not merely as a technical consideration but as a strategic driver of project success.

4.1.3 Definition and Scope of BLP

BLP method has been defined as a process of using buildability as a lens to evaluate elements/design options of a project to ensure that the most beneficial solution is selected from project conceptual planning, designing, and continuing across the project life. Instead of focusing solely on cost or design aesthetics, BLP encourages stakeholders to examine how design choices influence ease of construction, resource efficiency, safety, sustainability, and long-term performance.

The scope of BLP extends across the entire project lifecycle. In conceptual planning, BLP helps shape project strategies by identifying construction-friendly approaches, such as modularisation, standardisation, or simplified structural systems.

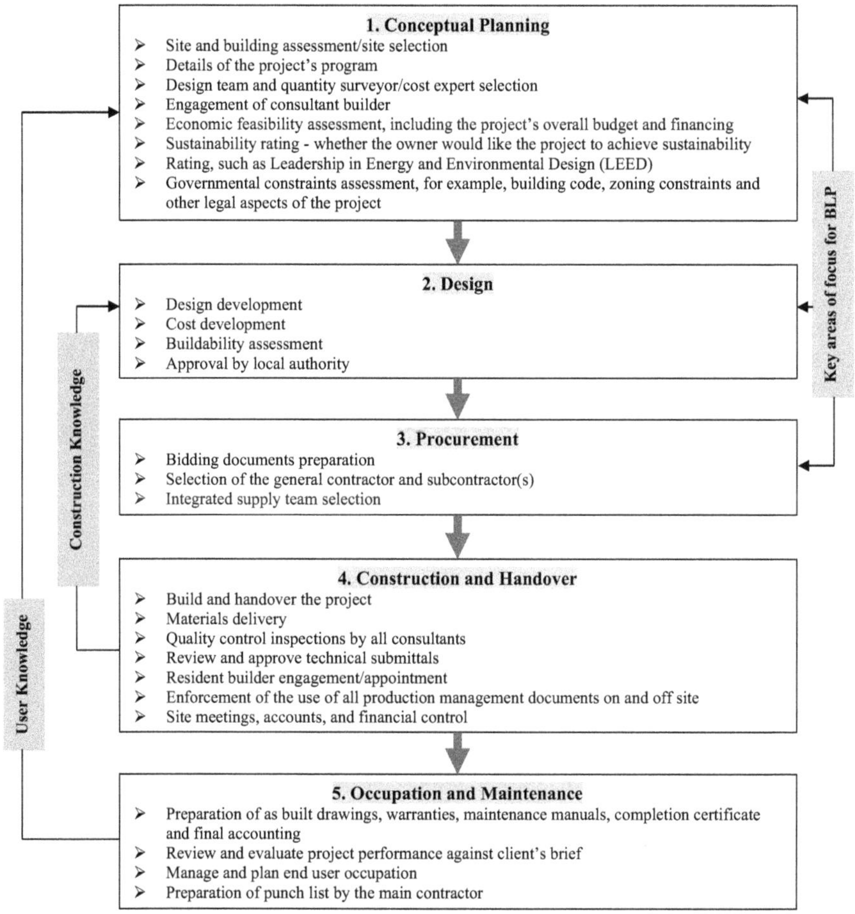

Figure 4.1 Visual representation of the conceptual framework for the BLP model.

Source: Osuizugbo (2025a, 2025b).

During detailed design, BLP applies constructability tools and methods to assess potential conflicts, risks, and inefficiencies. During construction, the procurement model facilitates continuous collaboration, enabling design adjustments where necessary and supporting innovation. In the operation and maintenance phase, BLP ensures that facilities are easy to manage, maintain, and adapt over time, enhancing asset lifecycle value.

BLP can be understood not only as a procurement method but also as a holistic project delivery framework that places buildability at the centre of all strategic, design, and operational decisions. Its definition and scope become clearer when examined alongside the visual diagrams representing its underlying logic and procedural flow. Figures 4.1 and 4.2, therefore, play a critical role in illustrating how BLP is conceptualised and how it functions in practice.

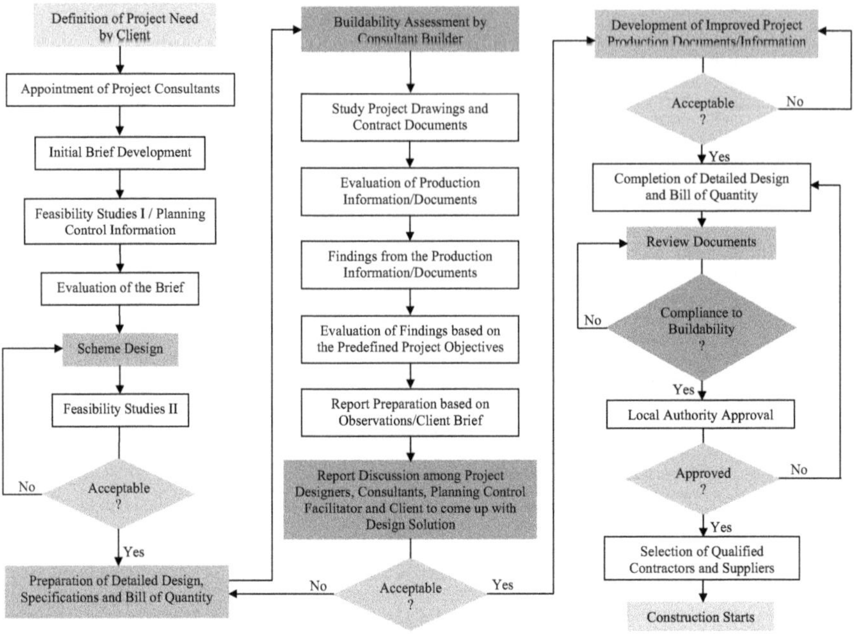

Figure 4.2 Buildability-led procurement model.

Source: Osuizugbo (2025a, 2025b).

Figure 4.1 provides the theoretical foundation that defines the identity and scope of BLP. It illustrates how the model is built upon the interaction of user knowledge (client-driven functional requirements, performance expectations, and operational needs) and construction knowledge (technical expertise, constructability logic, material understanding, sequencing, cost implications, and practical constraints). These dual knowledge streams form the intellectual basis for BLP, showing that procurement decisions must arise from a synthesis of client aspirations and professional construction inputs.

The conceptual framework identifies the key elements that BLP integrates, namely:

- Buildability – ensuring that design and construction processes are logical, practical, and efficient.
- Maintainability – ensuring that completed assets are easy to operate and maintain.
- Collaboration – facilitating open communication, teamwork, and integration of professional expertise.
- Supply Chain Integration – aligning contractors, subcontractors, and suppliers early in the process.
- Risk and Value Management – linking constructability decisions to risk reduction and value creation.

Conceptually, these components interact to form a procurement system that is responsive, adaptable, and performance-oriented. The framework likely depicts a flow from early conceptual planning, where client needs, regulations, and feasibility assessments are established, through design development, buildability evaluation, procurement, construction, and post-construction occupation. Each stage is influenced by the overarching goal of enhancing buildability, reducing waste, and improving project outcomes. Thus, Figure 4.1 serves as a theoretical map linking the principles that justify and shape the BLP model.

In essence, Figure 4.1 provides a theoretical grounding for the BLP model by showing how diverse project inputs, stakeholders, and processes converge to produce a buildability-driven procurement pathway. It underscores the holistic nature of BLP and emphasises that buildability must be integrated from project inception to completion rather than being treated as an isolated or late-stage consideration.

Furthermore, Figure 4.2 operationalises the conceptual framework by translating the theoretical components of BLP into a structured workflow. It shows the practical scope of BLP across five major phases: conceptual planning, design, procurement, construction and handover, and occupation and maintenance. Each phase contains clearly defined tasks, decision points, and feedback loops designed to ensure that buildability is assessed, validated, and enforced before the project progresses to the next stage.

1 *Conceptual Planning*: The process begins with site and building assessment, site selection, and the establishment of the project's overall requirements. Critical early tasks include defining the project team, designers, quantity surveyors or cost experts, and engaging a consultant builder, who plays a central role in BLP by providing professional buildability expertise. The conceptual planning stage also includes economic feasibility, budgeting, and the evaluation of sustainability ambitions such as Leadership in Energy and Environmental Design certification. Constraints from planning authorities, zoning, and legal frameworks are also assessed. This phase ensures that the project foundation aligns with buildability-driven principles before major commitments are made.
2 *Design Stage*: In the design phase, the model emphasises design development, cost development, and, explicitly, buildability assessment. The involvement of the consultant builder ensures that design decisions are analysed for constructability, efficiency, and sequencing. The design outputs undergo review and approval by local authorities. This phase highlights iterative loops: if production information is not acceptable, it is returned for revision. This ensures that constructability issues are resolved during design rather than emerging during construction, where they are more costly.
3 *Procurement Stage*: The procurement stage focuses on preparing bidding documents, selecting qualified contractors and subcontractors, and forming an integrated supply team. This integration is a key differentiation of BLP compared to traditional procurement, which typically isolates design from construction. By selecting supply-side partners early and based on capability rather than lowest

cost alone, the model encourages seamless collaboration and ensures that the construction team can fully participate in buildability refinement.

4 *Construction and Handover*: The construction phase includes critical activities such as appointing the resident builder, approving technical submittals, enforcing production management documents, conducting quality control inspections, and managing site meetings and financial controls. These steps illustrate the operational importance of the BLP model, production management, and buildability considerations are not abandoned once construction begins; rather, they are continually enforced and monitored. Materials delivery, inspection, and collaboration among consultants maintain the buildability ethos through construction until handover.

   Following completion, activities such as preparing punch lists, issuing warranties, and creating as-built documentation ensure that the project transitions smoothly to the occupation stage.

5 *Occupation and Maintenance*: BLP extends beyond construction to include performance evaluation and lifecycle considerations. The model ends with managing end-user occupation, evaluating project performance against the client brief, and providing maintenance manuals and completion certificates. This inclusion of post-construction activities reflects the BLP model's emphasis on maintainability, a core theme in BLP thinking. By linking procurement decisions to long-term asset performance, the model ensures that early buildability decisions translate into operational benefits.

Decision Loops and Compliance Gates: A distinctive feature of Figure 4.2 is the continuous evaluation loops and decision gates embedded throughout the process:

- Evaluation of the brief
- Scheme design review
- Feasibility Studies I and II
- Production information evaluations
- Local authority approvals
- Assessment of compliance with buildability
- Review of production documents

At each gate, the decision is binary: "Acceptable?" or "Not acceptable." If not acceptable, the workflow loops back to the earlier stage for correction. This ensures rigorous quality assurance and prevents premature progression to subsequent phases. It also enforces buildability compliance as a non-negotiable criterion, eliminating inefficiencies and constructability defects through iterative refinement.

Integration of Consultant Builder and Key Knowledge Inputs: The model highlights the central role of the consultant builder, whose expertise informs design and procurement decisions. Likewise, the dual inclusion of user knowledge and construction knowledge as "key areas of focus" reinforces that BLP is fundamentally collaborative. The client provides functional requirements, while the builder contributes construction logic; together, they optimise design and procurement decisions.

Overall Reflection on Figure 4.2: Figure 4.2 presents a comprehensive, step-by-step procurement model in which buildability acts as the primary organising principle. It replaces linear procurement models with a more integrated, circular, and evaluative framework. This ensures that constructability is optimised not only at the design stage but also throughout procurement, construction, and facility occupation. By embedding iterative assessments, expert involvement, and multidisciplinary collaboration, the model provides a structured method for achieving the goals of BLP: efficiency, cost-effectiveness, risk reduction, sustainability, and improved project outcomes.

Integrative Interpretation: Together, Figures 4.1 and 4.2 demonstrate that BLP is a comprehensive procurement philosophy supported by a clearly articulated operational system. The conceptual framework (Figure 4.1) defines what BLP is: a buildability-driven, knowledge-integrated, collaboratively structured procurement model. The procedural model (Figure 4.2) defines how BLP is executed: through phased processes, iterative evaluations, professional involvement of construction expertise, and structured decision-making checkpoints.

Through these visual representations, BLP emerges as a procurement method that transcends traditional boundaries between design and construction, establishing a continuous feedback-driven workflow aimed at achieving efficient, buildable, and maintainable project outcomes across the entire lifecycle of the built asset.

Table 4.1 provides a comparative assessment of the traditional procurement method versus the BLP approach, offering a clearer understanding of their distinctions and illustrating the enhanced effectiveness of the BLP method.

Table 4.1 Comparative analysis of traditional method and BLP method

No.	*Features*	*Traditional Method*	*BLP Method*
1	Collaboration and communication	Limited collaboration between design and construction teams.	Early involvement of builders improves collaboration and communication for buildability enhancement.
2	Procurement approach	Linear and chronological procurement process with separate contracts for construction and design.	Emphasises buildability assessment and enhancement, involving builders early in the design stage.
3	Decision-making authority	Decisions are made by individual parties based on contractual obligations.	Collaborative decision-making involving key stakeholders, with a focus on buildability enhancement.
4	Cost management	Fixed-price contracts may lead to cost overruns if unanticipated issues surfaces	Shared financial risk and early collaboration can lead to better cost control and value engineering.
5	Contract structure	Separate contracts for construction and design.	Contracts structured to prioritise buildability enhancement and collaboration.

(*Continued*)

Table 4.1 (Continued)

No.	*Features*	*Traditional Method*	*BLP Method*
6	Client involvement	Limited client involvement after the design stage.	Client involvement is continuous, promoting a collaborative and client-centric approach. It is vital, particularly in decision-making related to buildability implications.
7	Quality control	Quality control may be segmented, leading to potential gaps.	Emphasis on buildability enhancement should not compromise quality. Buildability-led allows for continuous quality improvement through collaboration.
8	Flexibility in design	Limited ability to adjust designs once contracts are in place.	Flexibility to adjust designs for buildability enhancement and value engineering.
9	Project flexibility	Limited flexibility once contracts are signed, with potential for change orders.	More adaptable to changes, allowing for continuous buildability assessment and amendment.
10	Risk allocation	Risks may be shifted to contractors, potentially resulting in adversarial relationships.	Shared risks among stakeholders and a focus on buildability enhancement can promote collaborative relationships and shared project goals.
11	Project timeline	Linear process may result in chronological stages, potentially extending project timelines.	Early collaboration can make processes more efficient, potentially minimising project timelines.
12	Project outcomes	Success may be measured individually by each part's adherence to contractual responsibilities.	Shared project risks, outcomes, and success create a more unified sense of accomplishment.

4.1.4 Expected Outcomes and Benefits of BLP

One of the strengths of BLP is the broad range of benefits it delivers to project owners, end-users, and the wider construction industry. Literature suggests that improved buildability correlates with enhanced project performance across time, cost, and quality dimensions. By incorporating structured buildability analysis and ECI, BLP is expected to lead to:

- Cost savings: Optimised design solutions reduce material waste, rework, and inefficiencies.

- Productivity improvement: Efficient design choices simplify construction tasks and sequencing.
- Design improvement: Early contractor input enhances the functionality, clarity, and practicality of design.
- Client satisfaction: Designs are more aligned with client needs and project specifications.
- Waste reduction: Rationalised construction methods reduce unnecessary material consumption.
- Improved communication: Collaborative culture strengthens relationships and reduces disputes.
- Innovation: Integrated teams stimulate creative solutions and adoption of modern methods.
- Value for money: Projects achieve better lifecycle performance due to enhanced durability and maintainability.

These benefits demonstrate that BLP is not solely about improving constructability but also about improving overall project value, stakeholder collaboration, and long-term asset performance.

4.1.5 The Role of Construction Expertise in BLP

A defining feature of BLP is the systematic involvement of construction experts, professional builders, construction managers, specialist contractors, and suppliers, throughout the procurement process. Their involvement ensures that design decisions are grounded in practical knowledge, including:

- feasibility of construction methods,
- material characteristics and limitations,
- sequencing and scheduling constraints,
- site logistics considerations,
- safety and environmental requirements,
- opportunities for modularisation or prefabrication,
- potential risks and mitigation options.

This professional insight helps identify design issues early, when they are least costly to address, thereby reducing downstream problems. It also facilitates innovation, as builders and suppliers can recommend alternative materials, construction technologies, or assembly techniques that can enhance buildability and performance.

The increase in project complexity has made this integration essential. As construction projects evolve to encompass advanced technologies, higher sustainability standards, and more demanding client requirements, designers alone cannot foresee all construction challenges. BLP's collaborative structure closes this gap, ensuring that decision-making is informed by the collective expertise of the project team.

4.1.6 Industry Drivers Necessitating BLP

The adoption of BLP has been driven by several industry challenges, given as follows:

- Fragmented procurement practices: Traditional procurement often isolates design from construction.
- Increasing project complexity: Large-scale and technologically intensive projects demand deeper integration.
- Sustainability pressures: Clients and regulations increasingly require resource-efficient, low-waste construction.
- Changing stakeholder needs: Modern clients demand higher quality, flexibility, and lifecycle value.
- Industry inefficiencies: Persistent cost overruns, rework, and delays highlight the need for improved procurement strategies.

These pressures have motivated governments, industry bodies, and private organisations to explore procurement models that promote better integration, accountability, and value generation.

BLP addresses these problems by creating a procurement environment that fosters collaboration, ensures early problem identification, and integrates innovative solutions into project delivery.

4.1.7 Design Flexibility and Evaluation Skills in BLP

A notable observation in the literature, particularly from Gorgolewski (2008), is the need for increased design flexibility and skilled evaluation of building components in the construction sector.

For BLP to function effectively, project teams must support:

- Flexible design specifications, allowing adjustments based on constructability insights.
- Skills in evaluating building components, including structural, environmental, and material considerations.
- Procurement systems that support component reuse and dismantling, reflecting sustainability and circular economy principles.

BLP responds to these requirements by embedding buildability analysis within the design process and by encouraging the adoption of standardised and modular design principles. This approach ensures that designs are not only efficient to construct but also adaptable, maintainable, and environmentally responsible.

4.2 ECI and Buildability

ECI is a collaborative procurement and delivery approach in which a contractor is engaged during the early design stages, often alongside designers, cost consultants, and the client, to contribute construction expertise before key decisions are locked in. The central objective of ECI is to improve buildability: the extent to which a project's design can be constructed efficiently, safely, economically, and to the required quality using available resources and methods.

4.2.1 Origin, Concept, and Rationale of ECI

The concept of ECI first appeared in formal literature through the Engineering and Construction Contract published in 1998 by the Institution of Civil Engineers (ICE) in England. Its early practical application is widely attributed to the United Kingdom's Road and Highways Agency, which began adopting ECI in the early 2000s. The primary aim at that time was to introduce buildability and constructability advice during the early stages of project development, rather than waiting until construction commencement.

Under this early implementation, contractors were selected based on capability and experience rather than based on the lowest price, enabling them to contribute meaningfully to design development. The perceived success of this approach led to the wider adoption of ECI after 2005 in several countries, including Australia, the United States, New Zealand, and the Netherlands. Since then, ECI has increasingly been recognised as an effective mechanism for improving collaboration and project outcomes in complex construction environments.

Traditionally, the design and construction phases of projects have been treated as separate and sequential processes. This separation, particularly under the design–bid–build (DBB) procurement approach, often results in buildability and constructability problems, as construction knowledge is largely excluded from early decision-making. Designers typically work closely with the client to produce solutions driven by budget, functionality, aesthetics, and engineering requirements. Contractors are then appointed through competitive tendering once the design is complete, leaving little or no opportunity for builders to influence design decisions.

Such traditional procurement approaches have been associated with inefficiencies, adversarial relationships, and limited integration of design and construction knowledge. In response to these shortcomings, ECI has emerged as an alternative project delivery approach that promotes early collaboration between the client, designer, and contractor. ECI refers to the involvement of the contractor during the early stages of project development, primarily to assist with planning, buildability, and constructability considerations.

The adoption of ECI is driven by several key objectives, including the integration of design and construction processes, improved value for money, enhanced collaboration, and more effective utilisation of market capacity. Unlike traditional procurement models, ECI is widely regarded as an innovative and proactive approach that emphasises collaborative working relationships and early risk management.

ECI is generally defined as an arrangement in which the main project parties, typically the client, designer, and contractor, engage during the early design stages of a construction project. ECI in design establishes a relationship that allows the contractor to contribute construction knowledge, experience, and practical expertise to the design process. In essence, ECI enables contractors to be engaged before the commencement of construction works in order to influence design outcomes positively.

Contractors often possess deeper practical knowledge of construction methods, materials, sequencing, pricing, and supply chains than designers or project owners. ECI leverages this expertise to achieve two primary aims: first, to enable contractors to make constructive and informed input into design development; and second, to allow clients to gain a clearer understanding of what is being procured and how the proposed design will meet business and operational needs.

As a project delivery strategy, ECI can be viewed as a hybrid approach that combines elements of traditional procurement methods such as DBB and design–build (DB) with relational models such as alliancing. By blending competitive principles with collaboration, ECI seeks to capture the advantages of both transactional and relational contracting. Consequently, ECI increasingly encourages long-term, cooperative relationships among project participants, positioning it as a relational contracting mechanism capable of addressing many of the persistent challenges facing the construction industry.

4.2.2 Buildability as a Core Objective of ECI

Buildability, sometimes referred to as constructability, describes the extent to which a project's design enables efficient, safe, and economical construction. It reflects how well design decisions take account of practical construction considerations, including methods, sequencing, resources, site conditions, and risk. A highly buildable design does not compromise architectural or functional intent; rather, it achieves project objectives while reducing unnecessary complexity and uncertainty during construction.

One of the key characteristics of a buildable design is the minimisation of complex or unnecessary detailing. Overly intricate details, while technically feasible, often increase construction time, cost, and the likelihood of errors on-site. Simplified and rationalised details improve clarity for contractors, reduce rework, and enhance quality by making it easier to achieve the design intent consistently. From a buildability perspective, simplicity is closely linked to reliability and efficiency.

Buildability is also enhanced through the use of standardised components and repeatable systems. Standardisation allows contractors to rely on familiar construction techniques, established supply chains, and proven installation processes. Repeatable elements improve productivity, reduce learning curves, and support prefabrication or modular construction, which can further improve quality and programme performance. Where repetition is embedded in the design, construction becomes more predictable and controllable.

Another critical aspect of buildability is the alignment of design intent with realistic construction methods. Designs that overlook how elements will actually be built, such as access for installation, tolerances, or sequencing, often lead to site-based design changes and inefficiencies. A buildable design considers how components will be assembled, in what order, and using which resources. This alignment ensures that construction methods are feasible within the project's time, cost, and labour constraints.

Consideration of site constraints, access, and logistics is equally central to buildability. Factors such as site size, location, surrounding infrastructure, ground conditions, and environmental limitations can significantly influence construction methods. Designs that acknowledge these constraints from the outset are more likely to facilitate efficient material delivery, safe movement of workers and equipment, and effective site organisation. Ignoring such constraints can result in congestion, delays, and increased safety risks during construction.

Safety is a fundamental dimension of buildability. Designs that reduce work at height, limit confined space activities, and allow for safer installation techniques contribute directly to improved health and safety outcomes. By considering construction hazards during design, buildable solutions can eliminate or reduce risks rather than relying solely on-site–based controls.

Through ECI, buildability is treated not as a late-stage compliance check, but as an integral design criterion. Contractors engaged during early design stages can challenge assumptions, identify impractical details, and propose alternative materials, systems, or layouts that achieve the required performance while improving constructability. This collaborative approach ensures that design solutions are informed by real construction experience, leading to more efficient, safer, and deliverable projects.

4.2.3 Timing and Forms of ECI

ECI can be implemented at varying levels of formality and contractual commitment, largely depending on the chosen procurement strategy, the scale and complexity of the project, and the client's risk appetite. Rather than a single fixed model, ECI represents a spectrum of approaches that range from informal advisory roles to fully integrated contractual arrangements. Each form offers different opportunities for influencing design and improving buildability.

One of the simplest forms is advisory ECI, where a contractor is engaged on a consultancy or professional services basis during the early design stages. In this arrangement, the contractor provides buildability and constructability reviews, cost planning advice, programme input, and guidance on construction methodology. However, the contractor has no contractual obligation or guaranteed right to deliver the construction works. Advisory ECI allows clients and designers to benefit from construction expertise while maintaining competitive tendering at a later stage. Its effectiveness depends heavily on the quality of advice provided and the willingness of the design team to incorporate contractor recommendations.

A more formal and widely adopted approach is two-stage procurement, in which a contractor is appointed early based on capability, experience, and preliminary pricing rather than a fully developed lump-sum price. During the first stage, the contractor collaborates with the client and designer to develop the design, refine construction methods, manage risks, and improve buildability. Once the design reaches sufficient maturity, a final price is negotiated for the second stage, leading to construction delivery. This approach balances early collaboration with cost certainty and is particularly effective for complex projects where design development benefits from contractor input.

At the most integrated end of the spectrum are integrated project delivery models or alliance contracts. Under these arrangements, the client, designer, and contractor work together under a single collaborative framework, often governed by shared risk and reward mechanisms. Commercial incentives are aligned with overall project performance rather than individual party outcomes. This high level of integration encourages transparency, joint decision-making, and collective ownership of risks and opportunities. As a result, buildability considerations are embedded throughout design development, rather than addressed in isolation.

Across all forms of ECI, timing is a critical factor. The earlier the contractor is involved, the greater the potential influence on fundamental design decisions that affect buildability, cost, programme, and safety. However, early involvement must be supported by clearly defined roles, responsibilities, and decision-making authority to avoid confusion and role overlap. When appropriately structured, varying forms of ECI provide flexible pathways for integrating construction expertise into design, thereby enhancing buildability and overall project performance.

4.2.4 Key Contributions of Contractors to Buildability

During ECI, contractors play a pivotal role in enhancing buildability by contributing practical knowledge derived from construction experience. Their early input enables design teams and clients to test design decisions against real-world construction constraints, thereby improving the likelihood that the final design can be delivered efficiently and safely.

One of the most significant areas of contractor contribution is construction methodology. Contractors can propose efficient construction sequencing, identify opportunities for prefabrication, and recommend off-site manufacturing strategies. Such approaches can reduce on-site labour requirements, improve quality control, and shorten construction durations. Early consideration of methodology ensures that the design supports, rather than restricts, efficient construction processes.

Contractors also add value through design simplification. Designs developed without construction input may include unnecessarily complex details that increase cost and risk without adding proportional value. Contractors can identify these issues early and suggest alternative details, materials, or systems that achieve the same functional, structural, or aesthetic objectives while being easier to construct. Simplified designs reduce the likelihood of errors, rework, and disputes during construction.

Temporary works and site logistics represent another critical contribution area. Contractors are uniquely positioned to advise on crane locations, lifting strategies, access routes, laydown areas, and site constraints that influence how permanent works should be designed. Early consideration of these factors can affect structural layouts, component sizes, and construction tolerances. By integrating temporary works and logistics planning into design development, projects can avoid costly late-stage modifications.

Programme optimisation is closely linked to buildability and benefits significantly from contractor input. Contractors can align design decisions with realistic construction durations, resource availability, and market conditions. This allows the design team to understand the programme implications of alternative design solutions and supports more accurate scheduling. A programme-informed design reduces the risk of unrealistic timelines and improves overall project certainty.

Finally, contractors contribute to risk identification and mitigation during ECI. Their experience enables early identification of buildability-related risks such as adverse ground conditions, complex interfaces between trades, tight tolerances, and supply chain constraints. Addressing these risks during design development allows for proactive mitigation strategies, rather than reactive problem-solving during construction.

Collectively, these contributions ensure that the final design is not only compliant with the client brief and regulatory requirements but also practical, efficient, and deliverable. By embedding contractor expertise early, ECI significantly enhances the overall buildability, constructability, and performance of construction projects.

4.2.5 Benefits of ECI for Project Outcomes

There are several benefits of ECI in construction projects that have been widely reported in the literature. ECI not only enables contractors to contribute directly to the design and planning process but also significantly improves communication and information flow between the project owner, designers, and contractors throughout both the design and construction phases. This enhanced interaction supports a more integrated project team and promotes shared understanding of project objectives, constraints, and risks.

From a client perspective, ECI delivers a broad range of benefits by leveraging the contractor's expert knowledge of construction processes to inform design development. Introducing construction knowledge during the early planning and design stages is critical, as decisions made at this point have the greatest influence on overall project cost, programme, and performance. By addressing buildability and constructability issues early, ECI minimises the likelihood of major design changes during construction, thereby reducing disruption and improving overall project efficiency.

One of the most significant benefits of ECI is improved cost certainty. Early contractor input allows potential cost drivers, market constraints, and construction risks to be identified and managed before designs are finalised. This reduces the risk of cost overruns arising from late design changes and provides the client with greater confidence in budget outcomes. Similarly, ECI contributes to improved programme performance by aligning design decisions with realistic construction durations, sequencing, and resource availability, leading to fewer delays and more reliable delivery schedules.

ECI also enhances quality outcomes. Designs informed by real construction practices are more likely to be built as intended, with fewer errors, reduced rework, and improved compliance with performance requirements. In addition, safety performance is improved through buildable designs that eliminate or reduce high-risk construction activities and allow safer construction methods to be planned during the design stage rather than managed reactively on-site.

Improved risk management is another key benefit of ECI. Early involvement enables contractors to identify project-specific risks related to ground conditions, materials, interfaces, access, and local site constraints. This proactive approach allows risks to be mitigated during design development rather than transferred or contested during construction. As a result, ECI often leads to fewer disputes and reduced adversarial behaviour, as risks and responsibilities are better understood and collectively managed.

ECI also provides tangible benefits for contractors. As the parties ultimately responsible for construction operations, contractors gain the opportunity to influence design decisions that affect constructability, cost, and programme. Early engagement allows contractors sufficient time to develop higher-quality construction planning and to provide project-specific advice on resource availability, performance limitations, and local practices. Continuous interaction between contractors and designers during design development further strengthens collaboration and leads to smoother relationships during construction.

Overall, the involvement of contractors in the early phases of planning and design provides an efficient and collaborative project delivery solution, while also improving client–contractor relationships. A clear understanding of these benefits, as summarised in Table 4.2, can further encourage the adoption of ECI as a preferred approach for construction project delivery.

Table 4.2 Benefits of ECI in construction projects

S/No	*Benefits*
1	Provides state-of-the-art knowledge from contractors
2	Improved buildability and constructability
3	It encourages innovation
4	Improved client–contractor relationship
5	Improved risk identification
6	Improved risk management
7	Improved safety
8	Life-cycle cost reduction
9	Dispute minimisation
10	Cost certainty
11	Cost saving
12	Reduced change orders
13	Schedule certainty
14	Improved time performance
15	Reduction in rework
16	Quicker start on-site

(*Continued*)

Table 4.2 (Continued)

S/No	*Benefits*
17	Improved contract documentation
18	Improved collaboration among parties
19	Improved quality performance
20	Good reputation is achieved
21	Better planning of resources requirements
22	Minimised environmental impacts
23	better forecasts of project outcomes
24	Facilitates good communication and understanding
25	Better site management
26	Attainment of efficient designs/improved drawing quality

4.2.6 Drivers for ECI in Construction Projects

Drivers for ECI can also be referred to as enablers or motivating factors that support and encourage the adoption of the ECI concept in construction project development. These drivers stem from long-standing challenges within the construction industry, including fragmented project delivery, poor buildability, cost overruns, schedule delays, and adversarial relationships. This section discusses the key drivers that have led to the increasing adoption of ECI in construction projects.

One of the primary drivers for ECI is the improvement in the quality of design outcomes. Involving contractors during the early stages of project development allows design decisions to be informed by practical construction knowledge. Contractors contribute insights into construction methods, material selection, sequencing, and site constraints, which enhance the feasibility and efficiency of design solutions. As a result, designs are more robust, coordinated, and aligned with construction realities.

Improved risk assessment and management is another significant driver for adopting ECI. ECI enables the identification of potential risks, such as ground conditions, constructability issues, interface risks, and supply chain constraints, during the planning and design phases. Since early-stage decisions have the greatest influence on project cost and outcomes, ECI provides an opportunity to mitigate risks proactively rather than reacting to them during construction.

The need for more realistic and reliable cost and schedule estimates also drives the adoption of ECI. Under traditional procurement approaches, budgets and programmes are often developed based on incomplete design information. Contractors possess up-to-date knowledge of market conditions, labour availability, material pricing, and construction durations. Their early input supports the development of more accurate cost plans and schedules, thereby improving price and scope certainty for clients.

Another important driver is the opportunity to enhance buildability, constructability, and innovation. Contractors can propose alternative construction methods, prefabrication strategies, and innovative technologies that may not be fully considered in design-led processes. Collaboration between designers and contractors creates an environment in which innovative solutions can be evaluated holistically in terms of performance, cost, time, and practicality.

Table 4.3 Drivers for ECI

S/No	*Drivers for ECI*
1	Price and scope certainty
2	Value engineering
3	Development of standard jointly
4	Client-contractor communication
5	Working relationship (Teamwork)
6	Innovation
7	Constructability
8	Productivity improvement
9	Transparency
10	Availability of relevant information
11	Buildability

Improvement in working relationships is also a key driver for ECI adoption. Frequent interaction between project parties during early project stages promotes better communication, trust, and mutual understanding. This collaborative environment reduces adversarial behaviour and supports effective teamwork throughout the project lifecycle. Improved client-contractor relationships often lead to smoother project delivery and fewer disputes during construction.

ECI is particularly advantageous in certain project scenarios. These include projects requiring immediate improvements to infrastructure, where early certainty and accelerated delivery are critical. ECI is also well suited to technically complex projects that are difficult to define at early stages, subject to change, or where multiple design alternatives require cost and schedule evaluation. Projects with high levels of coordination with external agencies, utilities, or stakeholders benefit from ECI due to early alignment and interface management. Additionally, projects that are highly sequence- or schedule-sensitive are well suited to ECI.

More specifically, three fundamental reasons underpin the selection of ECI as a project delivery approach. First, ECI improves the working relationship between the client and contractor through early engagement and collaboration. Second, it enhances buildability and constructability by incorporating the contractor's professional expertise into design development. Third, it increases opportunities for innovation through close collaboration between contractors and designers.

The drivers for ECI adoption are summarised in Table 4.3, which includes factors such as price and scope certainty, value engineering, development of joint standards, improved communication, teamwork, innovation, constructability, productivity improvement, transparency, availability of relevant information, and buildability. Understanding these drivers can further encourage the adoption of ECI as an effective approach to construction project development and delivery.

4.2.7 Challenges and Limitations of ECI

There is no single inherent disadvantage associated with ECI as a project delivery approach; however, despite its significant benefits, ECI faces a range of adoption and implementation challenges. These challenges do not negate the value of ECI, but they require careful management to ensure that the anticipated benefits are realised.

One of the most widely reported challenges in the literature is resistance to cultural change. Many organisations within the construction industry remain accustomed to traditional procurement models characterised by strict role separation, risk transfer, and adversarial relationships. Studies consistently identify resistance to culture change as the greatest barrier to ECI implementation. This resistance often stems from a lack of understanding of the ECI concept and its potential benefits. Improved awareness, training, and clear demonstration of successful ECI outcomes can significantly enhance acceptance and drive wider adoption.

Contracting practices and teamwork also present challenges. Early collaboration can blur traditional boundaries between designers and contractors, leading to role ambiguity and potential clashes over design ideas. If responsibilities, decision-making authority, and design leadership are not clearly defined, accountability may be undermined. Establishing clear governance structures, scopes of service, and communication protocols is therefore essential.

Another commonly perceived limitation relates to early-stage costs and remuneration. Engaging a contractor during the pre-construction phase requires additional upfront remuneration that would not typically be incurred under traditional procurement methods. Although this can create a perception of increased client cost, studies suggest that ECI can deliver savings many times greater than the initial contractor remuneration through reduced rework, improved buildability, and more efficient construction planning. Nevertheless, these early costs remain a commercial risk for the client, particularly where there is no guarantee that the contractor will ultimately agree to a fixed price acceptable to the client for construction delivery.

Commercial and pricing uncertainties represent further challenges. Due to limited design maturity at early stages, there may be insufficient time or information available to develop detailed or accurate pricing prior to commencement of works. This has implications for contractor selection criteria and for the processes used during the pre-construction phase to achieve acceptable levels of cost certainty before unconditional contractor appointment. Clients may also face difficulties in transferring risks that emerge later in the pre-construction stage if contractors are unwilling to accept them.

Additional commercial considerations include determining appropriate remuneration mechanisms for contractor activities during pre-construction, managing unequal levels of commitment between parties, and protecting commercially sensitive information shared during early collaboration. The open exchange of information, while beneficial for collaboration, can raise concerns regarding confidentiality and competitive advantage.

Collectively, these challenges highlight the need for careful structuring of ECI arrangements from the outset. As summarised in Table 4.4, understanding these limitations can facilitate the development of effective strategies to minimise risks associated with ECI implementation. When supported by transparent processes, clearly defined roles, and appropriate contractual frameworks, the challenges of ECI can be managed effectively, allowing its benefits to outweigh its limitations.

Table 4.4 Challenges of ECI implementation in construction projects delivery

S/No	*Challenges of ECI*
1	Lack of "win-win" attitude/common goal/team concept
2	Lack of commitment/support
3	Competitive tendering/traditional procurement
4	Inadequate training and guidance/experience/over-dependency
5	Lack of continuous, open, and honest communication
6	Cultural barriers/conservative approach
7	Lack of understanding/unwilling to compromise/suspicion
8	Lengthy selection/pre-contract time
9	Commercial pressures/insufficient efforts/role of facilitator
10	Bureaucratic organisations/lack of empowerment
11	Lack of technical knowledge/continuous improvement
12	Lack of trust/clear boundary/fear of the unknown
13	Lack of intimacy/ego or personality issues
14	Key parties not included
15	Failure to share risk/inappropriate allocation
16	Discreditable/past adversarial relationships/experience
17	Inefficient problem-solving/decision-making
18	Errors in drawings and specifications/contract language
19	Partnering is not suitable for all projects
20	Laws, regulations, rules, standard contracts, and legal issues
21	Inadequacy of remuneration
22	Unfamiliarity with ECI process
23	Change in relationship protocol

4.2.8 Best Practice Principles

Successful implementation of ECI for enhanced buildability depends on the application of several best practice principles that support collaboration, clarity, and value creation. These principles provide a practical framework for translating the concept of ECI into effective project delivery outcomes.

A fundamental principle is to engage contractors based on capability, experience, and collaborative behaviours rather than price alone. Because ECI focuses on early-stage value creation rather than immediate cost competition, contractor selection should emphasise technical competence, relevant project experience, problem-solving ability, and willingness to collaborate. Selecting contractors solely on the lowest price at this stage can undermine the quality of input and discourage innovation.

It is equally important to clearly define the scope and expectations of contractor involvement during the early stages. Ambiguity regarding roles, responsibilities, and decision-making authority can lead to confusion and conflict. A clear definition of the contractor's pre-construction services, such as buildability reviews, cost advice, risk identification, and construction planning, helps maintain accountability while enabling meaningful contribution.

Open information sharing and early risk discussions are central to effective ECI. Transparent communication allows project participants to identify constraints, uncertainties, and risks at a stage when they can be most effectively addressed. Early risk discussions promote collective problem-solving and reduce the likelihood of disputes arising later in the project. Trust and openness are therefore essential enablers of buildability-driven outcomes.

Another key principle is to align commercial incentives with overall project outcomes, particularly buildability, value for money, and performance. Commercial arrangements should encourage behaviours that support collaboration rather than cost shifting or risk avoidance. Mechanisms such as open-book costing, shared savings, or performance-based incentives can help ensure that all parties are motivated to optimise project outcomes rather than individual gains.

Finally, successful ECI requires maintaining clear design leadership while respecting construction expertise. Designers remain responsible for design intent, compliance, and quality, while contractors contribute practical construction knowledge to inform decision-making. Mutual respect between disciplines ensures that buildability improvements are achieved without compromising functional, aesthetic, or regulatory requirements.

When these best practice principles are consistently applied, ECI becomes a powerful mechanism for embedding buildability into the core of a project. Rather than treating constructability as a late-stage check, ECI enables buildability to be integrated into decision-making from the outset, resulting in safer, more efficient, and more predictable project delivery.

4.3 Buildability in Design

Buildability in design refers to the extent to which a project's design facilitates efficient, safe, economical, and high-quality construction. It reflects how effectively design decisions consider construction methods, sequencing, resources, site conditions, and practical constraints. Rather than focusing solely on compliance with functional and aesthetic requirements, buildability emphasises the integration of construction knowledge into the design process to ensure that what is designed can be delivered efficiently in practice.

Historically, the separation of design and construction under traditional procurement models has contributed to poor buildability outcomes. Designers often develop solutions based on technical standards, performance criteria, and architectural intent, while construction considerations are addressed only after designs are finalised. This disconnect frequently results in designs that are difficult to build, require extensive temporary works, or necessitate redesign during construction. Buildability in design seeks to overcome these limitations by embedding construction thinking into early decision-making.

A fundamental aspect of buildability is design simplicity and clarity. Overly complex details, excessive tolerances, or bespoke solutions can significantly increase construction time, cost, and risk. Simplified and standardised design solutions reduce ambiguity, minimise the potential for errors, and improve productivity on-site. Buildable designs favour repetition, modularity, and rationalised detailing, which allow contractors to apply proven construction techniques and achieve consistent quality.

Another critical dimension of buildability is the alignment of design intent with realistic construction methods. Designers must consider how components will be fabricated, transported, installed, and assembled. This includes accounting for lifting capacities, access requirements, sequencing constraints, and tolerances between trades. Designs that fail to consider these factors often lead to site-based improvisation, inefficiencies, and increased safety risks. A buildable design anticipates the construction process and supports logical, efficient workflows.

Site constraints and logistics also play a central role in buildability. Factors such as site size, location, access routes, surrounding infrastructure, ground conditions, and environmental restrictions directly influence feasible construction methods. Buildability-focused design acknowledges these constraints from the outset, ensuring that layouts, structural systems, and component sizes are compatible with site realities. Failure to consider logistics during design can result in congestion, delays, and costly modifications during construction.

Safety is an integral component of buildability in design. Many construction hazards originate from design decisions, such as excessive work at height, complex temporary works, or confined space requirements. Buildable designs aim to eliminate or reduce hazards through safer design choices, such as prefabrication, modular construction, and simplified installation processes. By addressing safety risks during design, reliance on-site–based controls is reduced, leading to improved health and safety outcomes.

Buildability is also closely linked to cost and programme performance. Early design decisions determine the majority of a project's cost and construction duration. Designs that incorporate buildability principles tend to require fewer changes during construction, experience less rework, and achieve more predictable schedules. Conversely, poor buildability often results in claims, delays, and budget overruns. From a client perspective, buildable design directly contributes to value for money and delivery certainty.

The integration of buildability in design is significantly enhanced through early collaboration, particularly through approaches such as ECI. Contractors possess specialised knowledge of construction methods, materials, productivity rates, and local practices that designers and clients may lack. When contractors are involved early, they can review design proposals, identify constructability issues, and suggest alternatives that maintain performance while improving efficiency. This collaborative process transforms buildability from a reactive review into a proactive design driver.

Despite its importance, achieving buildability in design presents challenges. Designers may be constrained by limited construction experience, tight design timelines, or rigid procurement processes. There may also be tension between architectural ambition and construction practicality. Overcoming these challenges requires a cultural shift toward collaboration, supported by procurement strategies, contractual frameworks, and governance structures that encourage early and meaningful input from construction specialists.

In contemporary construction projects, buildability in design has become increasingly critical due to rising project complexity, technological advancement, and sustainability requirements. Digital tools such as Building Information Modelling further support buildability by enabling clash detection, sequencing simulation, and constructability analysis during design development. When combined with collaborative delivery models, these tools enhance the ability to design with construction in mind.

In summary, buildability in design is a fundamental determinant of project success. It ensures that design solutions are not only technically and aesthetically sound but also practical, safe, and efficient to construct. By embedding construction knowledge into the design process, projects can achieve improved cost certainty, programme reliability, quality, and safety. As the construction industry continues to seek better performance and value, buildability in design remains a cornerstone of effective project delivery.

4.4 Buildability in Construction Practice

Buildability in construction practice refers to the practical application of buildability principles during the execution of construction works. While buildability in design focuses on embedding construction considerations into drawings and specifications, buildability in practice concerns how effectively those intentions are realised on-site through planning, coordination, sequencing, and management of construction activities. It represents the translation of design intent into physical reality and is therefore a critical determinant of project performance.

In construction practice, buildability is closely associated with construction planning and methodology selection. Contractors must determine how works will be executed, including sequencing of activities, allocation of resources, and selection of construction techniques. Projects with high buildability enable straightforward sequencing, efficient workflows, and minimal interference between trades. Conversely, poor buildability often results in congested sites, inefficient work sequences, and frequent disruptions, leading to productivity losses and delays.

A key aspect of buildability in practice is the management of interfaces between trades and work packages. Modern construction projects involve numerous specialist subcontractors whose activities are interdependent. Effective buildability allows these interfaces to be clearly defined, coordinated, and managed. When buildability is lacking, interface clashes occur, resulting in rework, disputes, and programme slippage. Construction managers play a crucial role in mitigating these issues through detailed coordination, method statements, and look-ahead planning.

Site logistics and material handling are central to buildability in construction practice. Decisions regarding access routes, storage areas, crane locations, lifting strategies, and traffic management have a direct impact on construction efficiency and safety. Buildable projects allow materials to be delivered, stored, and installed with minimal handling and disruption. Poor logistical planning, often stemming from inadequate buildability consideration, can cause congestion, increased labour effort, and heightened safety risks.

Another critical dimension of buildability in practice is temporary works design and management. Temporary structures such as scaffolding, formwork, shoring, and access platforms are essential to construction operations. Projects that are buildable in practice integrate temporary works considerations into construction planning, ensuring that they are safe, economical, and compatible with permanent works. Failure to adequately plan temporary works often leads to delays, redesign, or unsafe site conditions.

Health and safety performance is strongly influenced by buildability in construction practice. Many site hazards arise from complex construction processes, constrained access, or poorly sequenced activities. Buildable construction practices seek to eliminate or reduce risks through safer methods such as prefabrication, modularisation, mechanised installation, and reduced work at height. By adopting buildable construction methods, contractors can move beyond reliance on administrative controls and personal protective equipment toward inherently safer systems of work.

Buildability in practice also affects productivity and workforce efficiency. Clear work sequences, standardised components, and repeatable tasks enable crews to work more efficiently and with fewer errors. Buildable projects reduce uncertainty and learning curves, allowing workers to focus on execution rather than problem-solving. In contrast, poor buildability increases cognitive and physical demands on workers, leading to fatigue, mistakes, and reduced productivity.

Quality management is another area where buildability plays a significant role in construction practice. Buildable designs and construction methods allow work to be executed accurately and consistently, reducing the likelihood of defects and rework. When construction processes align with design assumptions, achieving specified tolerances and performance requirements becomes more feasible. Poor buildability often forces site-based improvisation, which compromises quality and increases the risk of non-conformance.

The role of communication and collaboration is particularly important in achieving buildability during construction. Effective information flow between designers, contractors, subcontractors, and suppliers enables issues to be resolved quickly and constructively. Approaches such as ECI and collaborative contracting enhance buildability in practice by ensuring that construction challenges identified on-site can be addressed through coordinated decision-making rather than adversarial processes.

Buildability in construction practice is also influenced by resource availability and market conditions. Labour skills, material supply chains, equipment availability, and local regulations all affect how buildable a project is in practice. Contractors must adapt construction methods to reflect these realities, reinforcing the importance of flexible planning and early engagement with suppliers and specialists.

Despite its importance, achieving buildability in construction practice is not without challenges. Projects may inherit poor buildability from design stages, leaving contractors to manage inefficiencies during execution. Time pressures, contractual constraints, and fragmented responsibilities can further limit opportunities to improve buildability once construction has commenced. These challenges highlight the need for strong integration between design and construction processes.

In summary, buildability in construction practice is a critical factor influencing cost, time, quality, safety, and productivity on construction projects. It reflects how well construction activities are planned, coordinated, and executed in alignment with design intent and site realities. When buildability is prioritised in practice, projects are more likely to achieve efficient workflows, safer working conditions, and predictable outcomes. As construction projects continue to increase in complexity, effective management of buildability in construction practice remains essential for successful project delivery.

4.5 Maintainability and Buildability

Maintainability and buildability are two interrelated concepts that play a vital role in engineering, construction, and infrastructure development, as they directly affect a project's long-term performance, cost efficiency, and sustainability. Buildability refers to the extent to which a project's design facilitates ease of construction using available resources, skills, technologies, and construction methods. A buildable design minimises construction difficulties, reduces errors, shortens project timelines, and lowers initial construction costs. It focuses on practical aspects such as construction sequencing, material handling, workforce efficiency, and site constraints.

Maintainability, on the other hand, concentrates on the operational phase of a facility after construction is complete. It addresses how easily systems and components can be accessed, inspected, serviced, repaired, or replaced throughout the asset's life cycle. Good maintainability reduces downtime, enhances safety for maintenance personnel, and limits long-term operational and repair costs. Design considerations such as accessibility, durability of materials, simplicity of systems, and use of standardised components are central to achieving high maintainability.

When buildability and maintainability are considered together from the early design and planning stages, projects benefit significantly. Design choices that simplify construction while also allowing easy future access and maintenance help prevent costly modifications later. Early integration of these concepts encourages collaboration among designers, contractors, and facility managers, leading to better decision-making. Ultimately, balancing buildability and maintainability results in higher-quality infrastructure that is not only efficient to construct but also economical, reliable, and sustainable to operate over its entire life span.

4.5.1 Overview of Maintainability

Maintainability refers to the ease, safety, and efficiency with which a system, structure, or facility can be inspected, serviced, repaired, and upgraded throughout its operational life. It is a critical design and management consideration that directly influences reliability, availability, operating costs, safety, and overall value of built assets. A maintainable facility allows routine and corrective maintenance activities to be carried out with minimal disruption, reduced labour effort, and efficient use of resources, while consistently meeting performance and safety requirements. Integrating maintainability into the early design and planning stages is therefore essential for reducing downtime, preventing avoidable failures, and optimising life-cycle costs.

In the construction process, project performance is traditionally evaluated based on the equilibrium between time, cost, and quality. Clients typically desire facilities that are delivered quickly, at the lowest possible cost, and with the highest achievable quality. The responsibility for balancing these three factors is shared among the project team, including architects, engineers, contractors, and project managers. However, quality should not be limited to visual appearance or initial performance alone; it must also encompass long-term functionality, ease of maintenance, operational efficiency, and sustainability. A building that is inexpensive and fast to construct but costly and difficult to maintain cannot be considered truly successful.

Better building design inevitably leads to better buildings and greater value for owners. However, defining "better design" is complex, as buildings are judged from multiple perspectives, including aesthetics, functionality, occupant comfort, return on investment, constructability, maintainability, and environmental performance. Modern buildings are increasingly complex, incorporating advanced structural systems, mechanical and electrical services, sustainability technologies, and digital controls. Their successful design, construction, and operation require the involvement of many participants, such as architects, engineers, contractors, financiers, project managers, and specialist consultants in areas like fire safety, acoustics, and environmental impact. In this context, an integrated team approach is essential. Fragmented design processes, where consultants work in isolation with limited coordination, often result in poor maintainability outcomes, as key decisions affecting long-term performance are made without adequate consideration of maintenance implications.

The concept of maintainability was formally introduced by the United States military in 1954, with the primary objective of improving maintenance efficiency and effectiveness. In engineering terms, maintainability is defined as the ease with which a product or system can be maintained to identify faults, correct defects, adapt to new requirements, and operate effectively under changing conditions. In the context of buildings, maintainability encompasses all processes and practices that ensure a facility can be economically maintained, repaired, and upgraded throughout its life span. Well-maintained buildings not only last longer but also provide safer, healthier, and more comfortable environments for occupants.

One of the most important aspects of maintainability is accessibility. Building components that require frequent inspection, cleaning, adjustment, or replacement, such as HVAC systems, electrical panels, plumbing networks, and façade elements, must be easily accessible. Poor access increases maintenance time, labour costs, and safety risks, often leading to deferred maintenance and premature failures. Providing adequate working space, safe access routes, removable panels, gangways, and clear labelling enables maintenance tasks to be carried out efficiently and safely, particularly in large or high-rise buildings where access challenges are more pronounced.

Simplicity of design is another fundamental principle of maintainability. Simple systems with fewer components and clear functional layouts are easier to understand, diagnose, and repair. Overly complex designs increase the likelihood of faults, complicate maintenance procedures, and demand highly skilled labour and specialised equipment. As building complexity increases, so do maintenance costs. Simple and logical design solutions reduce human error, shorten training time for maintenance personnel, and enhance overall reliability by limiting potential points of failure.

Standardisation significantly improves maintainability by ensuring that components, materials, and systems are interchangeable and readily available. Standardised parts reduce spare inventory requirements, shorten repair times, and lower procurement costs. They also allow maintenance teams to develop familiarity and expertise, improving efficiency and consistency in maintenance operations. In contrast, customised or non-standard components often result in long delays and higher costs when replacements are required.

Material selection also plays a critical role in maintainability. Durable materials that resist corrosion, wear, fatigue, and environmental degradation require less frequent maintenance and replacement. Although such materials may increase initial construction costs, they typically result in substantial long-term savings by reducing maintenance interventions and extending service life. Inappropriate material choices, such as materials unable to withstand local climate conditions or incompatible materials used together, are a common source of maintenance problems.

Safety considerations are integral to maintainability. Maintenance activities should be designed to minimise risks by providing safe access, adequate lighting, proper ventilation, and clear isolation of services. Unsafe maintenance conditions not only endanger personnel but can also lead to operational delays, legal liabilities, and regulatory non-compliance.

Effective maintainability planning is a collective responsibility that requires collaboration among all stakeholders, from brief development through design, construction, and operation. Maintainability analysis should begin early in the design phase, where potential maintenance problems are identified and addressed before construction. This analysis requires sound knowledge of building materials, construction technology, building services, structural behaviour, and costing. Designers must ask critical questions regarding access, dismantling, replacement frequency, availability of spare parts, working space, cleaning methods, and safety provisions.

Failure to incorporate maintainability considerations during design often leads to common deficiencies, such as inadequate space around equipment, poor building orientation, inappropriate material selection, excessive focus on aesthetics, and complex forms that increase maintenance costs. High-rise and complex-shaped buildings, for example, generally incur higher maintenance costs due to increased services, access difficulties, and temporary support requirements.

In conclusion, maintainability is a key parameter in optimising building availability, performance, and life-cycle cost. High maintainability can only be achieved through early, coordinated efforts that integrate design, construction, and operational perspectives. When maintainability is treated as a core design objective rather than an afterthought, buildings deliver better value, improved performance, enhanced safety, and long-term sustainability for owners and users alike.

4.5.2 Nexus between Buildability and Maintainability

Buildability and maintainability are closely interconnected concepts that significantly influence the overall success and long-term performance of construction projects. Decisions made during the planning, design, and construction stages not only affect how easily and efficiently a facility can be built but also determine how effectively it can be operated, maintained, and adapted throughout its service life. Buildability focuses on the efficiency of the construction process by considering available resources, construction methods, labour skills, site conditions, and sequencing. Maintainability, in contrast, emphasises the ease, safety, and cost-effectiveness of inspection, repair, replacement, and operation after the facility is completed. When buildability is prioritised without adequate consideration of maintainability, projects may achieve short-term construction efficiency but incur long-term operational difficulties and high maintenance costs.

The nexus between buildability and maintainability is first evident in design decisions. Designs that are simple, well-coordinated, and construction-friendly often enhance maintainability as well. For example, layouts that allow straightforward construction sequencing can also provide clear access routes for future maintenance. Conversely, designs that aim solely to simplify construction, such as tightly packed services or concealed components, may restrict access and complicate future repairs. Therefore, design decisions should balance ease of construction with long-term access, replacement, and adaptability to ensure life-cycle efficiency.

Construction methods also play a critical role in linking buildability and maintainability. Methods that promote accuracy, quality workmanship, and proper installation reduce defects that commonly lead to early failures and frequent maintenance. Poor workmanship during construction, even if it accelerates project completion, often results in hidden faults that manifest during operation. Prefabrication and modular construction, when properly designed, can enhance both buildability and maintainability by improving quality control and allowing easier replacement of components.

Material selection further illustrates the interdependence of buildability and maintainability. Materials chosen for ease of handling, installation, or reduced construction time must also satisfy durability, resistance to environmental conditions, and maintenance requirements. While lightweight or innovative materials may improve buildability, they may perform poorly over time if durability and maintenance needs are overlooked. Selecting materials that balance constructability with long-term performance reduces repair frequency and operational costs.

The integration of services, including mechanical, electrical, and structural systems, is another critical area where buildability and maintainability intersect. Poor coordination during construction can result in congested service spaces, restricted access, and conflicts between systems, making future maintenance difficult and unsafe. Effective integration ensures that systems are installed logically, with sufficient space and access for inspection, servicing, and upgrades.

A well-balanced approach that integrates buildability and maintainability ensures that a project is not only economical and efficient to construct but also sustainable, safe, and practical to operate over its lifespan. Achieving this balance requires early and continuous collaboration among designers, constructors, and maintenance professionals. By considering construction and maintenance requirements simultaneously, project teams can deliver facilities that provide long-term value, reduced life-cycle costs, and improved performance for owners and users alike.

4.6 Overview of Construction Methodology

Construction methodology refers to the systematic and organised approach adopted to plan, manage, and execute construction activities in order to transform design concepts into completed physical facilities. It defines how a project will be built and includes decisions related to construction techniques, sequencing of work, selection of equipment and materials, labour deployment, safety provisions, quality control measures, and site logistics. An effective construction methodology provides a clear roadmap for project execution and ensures that construction activities are carried out efficiently, safely, and in compliance with design specifications, regulatory requirements, and contractual obligations.

In the contemporary construction environment, projects are becoming increasingly complex due to advanced building technologies, sustainability requirements, tighter schedules, and rising cost pressures. As a result, construction methodology is no longer treated as a concern limited to the execution phase alone. Instead, it is closely linked to the early design and planning stages, where decisions have a profound impact on constructability, cost certainty, and project duration. Poorly defined or inappropriate construction methodologies often result in delays, rework, inefficient use of resources, safety incidents, and compromised quality. Conversely, a well-planned methodology enhances productivity, minimises waste, reduces risk, and contributes to overall project success.

Construction methodology also plays a vital role in balancing the traditional project objectives of time, cost, quality, and safety. It provides a structured framework within which these objectives can be achieved simultaneously. By clearly defining construction processes and responsibilities, the methodology facilitates coordination among different trades and stakeholders, improves communication on-site, and supports effective decision-making throughout the construction phase.

4.6.1 Conceptual Link: Buildability and Construction Methodology

Buildability is both conceptually and practically linked to construction methodology, as it represents the bridge between design intent and on-site execution. While design defines what is to be built, construction methodology determines how it will be built. Buildability focuses on the extent to which a design facilitates efficient construction using available resources, labour skills, technologies, and methods, without compromising functional, structural, aesthetic, or performance requirements. Construction methodology, in turn, is the practical mechanism through which buildability principles are translated into action on-site. Together, they form an integrated framework that directly influences project efficiency, quality, safety, and overall success.

At a conceptual level, buildability acknowledges that design decisions have a profound impact on construction processes. Choices related to structural systems, building form, materials, detailing, and layout determine the complexity of construction operations. A design that is buildable allows construction activities to be carried out logically and efficiently, minimising unnecessary effort, waste, and risk. Construction methodology responds to these design choices by selecting appropriate construction techniques, sequencing strategies, equipment, and resource allocations that align with the design's characteristics. When design and methodology are well aligned, construction becomes more predictable, controllable, and economical.

A strong conceptual link exists because construction methodologies are directly shaped by the buildability of the design. Designs that incorporate buildability principles tend to favour simplicity, clarity, repetition, and standardisation. For example, repetitive floor layouts, standardised structural grids, and uniform detailing enable the use of consistent construction methods across the project. This allows contractors to develop streamlined methodologies, reduce learning curves, and improve productivity. Workers become familiar with repetitive tasks, which enhance speed, accuracy, and quality. Standardisation also supports the use of prefabrication and modular construction techniques, which further improve efficiency and reduce on-site risks.

Clear and logical structural systems also strengthen the link between buildability and construction methodology. Designs that clearly define load paths, structural connections, and construction tolerances enable straightforward construction planning and execution. Construction methodologies can then be developed with confidence, using conventional techniques and readily available equipment. In contrast, overly complex or unconventional designs often require specialised construction methods, bespoke formwork, intricate temporary works, or highly skilled labour. While such designs may achieve aesthetic or symbolic goals, they frequently increase construction time, cost, and risk if buildability is not adequately considered.

When buildability is neglected during design, construction methodology becomes reactive rather than proactive. Contractors are forced to devise complex methodologies to overcome design constraints, such as restricted access, congested service spaces, or non-standard components. This often results in inefficient sequencing, increased reliance on temporary works, and greater coordination challenges among trades. Difficult access arrangements can compromise safety and slow progress, while intricate sequencing increases the likelihood of errors and rework. Consequently, projects may experience delays, cost overruns, quality issues, and safety incidents. These outcomes highlight the importance of embedding buildability considerations into design decisions that directly inform construction methodology.

The integration of buildability into construction methodology is most effective when it occurs early in the project lifecycle. Early-stage collaboration among designers, engineers, contractors, and construction managers is essential for identifying potential construction challenges and addressing them before construction begins. This collaborative approach allows construction expertise to inform design decisions, ensuring that the proposed design can be efficiently constructed using available resources and methods. ECI, constructability reviews, and design workshops are common mechanisms for facilitating this integration.

Through early collaboration, project teams can evaluate alternative design options from a construction perspective, comparing their impacts on methodology, cost, time, and risk. For example, alternative structural systems can be assessed based on ease of construction, availability of materials, and compatibility with site constraints. Construction sequencing can be considered alongside design development, ensuring that the design supports logical and efficient workflows. Equipment requirements, access routes, and site logistics can also be incorporated into design decisions, reducing the need for costly adjustments during construction.

Aligning design intent with practical construction considerations enables the selection of appropriate construction techniques, equipment, and sequencing strategies. This alignment reduces uncertainty and improves planning accuracy, allowing construction methodologies to be developed in a systematic and coordinated manner. As a result, rework is minimised, as potential conflicts and constructability issues are resolved at the design stage rather than on-site. Quality is improved because construction activities are carried out under controlled and well-planned conditions, and safety is enhanced through the reduction of hazardous and improvised operations.

The conceptual link between buildability and construction methodology also extends to sustainability and life-cycle performance. Efficient construction methodologies derived from buildable designs reduce material waste, energy consumption, and environmental impact during construction. They also support better workmanship and quality, which contribute to improved durability and reduced maintenance requirements during operation. In this way, the integration of buildability and construction methodology supports not only construction efficiency but also long-term project value.

In summary, buildability and construction methodology are inseparable elements of successful project delivery. Buildability provides the guiding principles that shape design decisions, while construction methodology translates those principles into practical actions on-site. When design and methodology are aligned through early collaboration and integrated planning, projects benefit from improved efficiency, reduced risk, enhanced quality, and better overall outcomes. Recognising and strengthening the conceptual link between buildability and construction methodology is therefore essential for delivering complex construction projects that meet client objectives and perform effectively throughout their life cycle.

Figure 4.3 illustrates how buildability serves as a crucial link between design decisions and construction methodology. The process begins with the establishment of project objectives, which define the expectations for time, cost, quality, and safety. These objectives then guide design decisions, encompassing aspects such as building form, layout, materials, and detailing. Buildability assesses whether these design decisions can be efficiently constructed using the available resources, labour skills, and technologies. A design that incorporates buildability principles directly

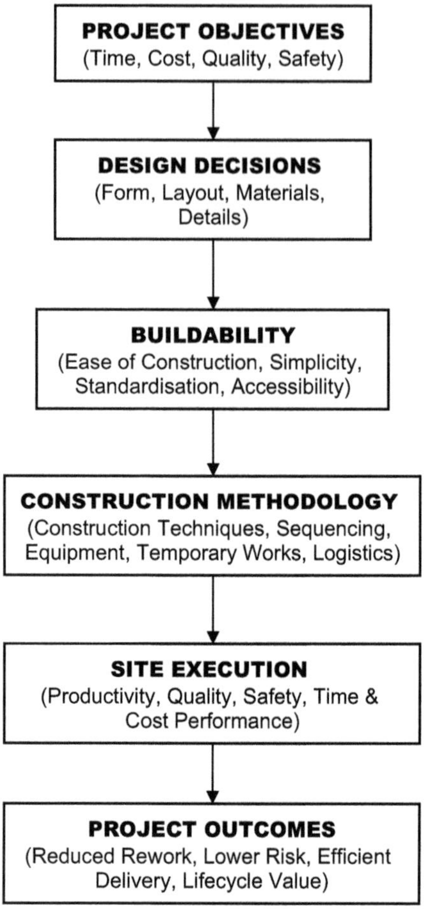

Figure 4.3 Conceptual link between buildability and construction methodology.

influences the construction methodology by determining appropriate construction techniques, sequencing of activities, selection of equipment, and site logistics. The chosen methodology, in turn, affects site execution, impacting productivity, safety, quality, and overall project performance. Ultimately, this integration of buildability into construction methodology results in improved project outcomes, including reduced rework, minimised risks, and enhanced lifecycle value of the facility.

4.6.2 Key Elements of Buildability in Construction Methodology

Buildability plays a crucial role in shaping construction methodology and ultimately determines how efficiently, safely, and economically a project can be delivered. It ensures that design decisions translate into practical construction processes that make effective use of available resources, technologies, and skills. The incorporation of buildability into construction methodology is achieved through several key elements, each of which directly influences project performance in terms of time, cost, quality, safety, and sustainability.

Design Simplicity and Clarity: Design simplicity and clarity form the foundation of buildability in construction methodology. Simple, well-coordinated designs reduce ambiguity and minimise construction complexity, allowing contractors to plan and execute work efficiently. Clear and consistent drawings, specifications, and details enable construction teams to understand design intent without excessive interpretation, reducing the likelihood of errors, miscommunication, and rework. Overly complex designs with unnecessary variations, intricate detailing, or poorly coordinated drawings often require complex construction techniques and highly skilled labour, increasing both cost and risk. In contrast, simple and logical design layouts allow construction methodology to be straightforward, with predictable workflows and minimal disruption. Design clarity also supports accurate quantity take-offs, scheduling, and cost estimation, which are essential for effective construction planning. By emphasising simplicity and clarity, buildability ensures that construction methodology is efficient, reliable, and adaptable to site conditions.

Construction Sequencing and Planning: Construction sequencing and planning are critical elements through which buildability is realised on-site. Buildable designs facilitate logical and efficient sequencing of construction activities, ensuring that tasks are performed in an orderly manner with minimal interference between trades. Proper sequencing reduces congestion, avoids conflicts, and improves overall workflow, contributing to higher productivity and shorter construction durations. When design supports buildable sequencing, construction methodologies can be developed to optimise resource utilisation and minimise downtime. For example, repetitive structural layouts enable floor-by-floor or zone-based construction sequences that are easy to manage and control. Poorly considered designs, however, may force contractors into inefficient sequencing, such as installing services in confined spaces after finishes are completed, leading to rework and delays. Effective sequencing, supported by buildability principles, enhances coordination among trades and ensures the smooth progression of construction activities.

Standardisation and Repetition: Standardisation and repetition are key drivers of efficiency in construction methodology. The use of standardised components, dimensions, and details allows construction teams to adopt repetitive construction processes that improve speed, accuracy, and quality. Repetition reduces the learning curve for workers, enabling them to perform tasks more efficiently as they become familiar with the methods and components involved. Standardisation also supports mass production and prefabrication, which can significantly reduce construction time and cost. Uniform details and components simplify procurement, reduce material waste, and minimise the risk of installation errors. From a construction methodology perspective, standardised designs enable the use of consistent tools, equipment, and methods across the project, improving coordination and control. This approach is particularly beneficial in large-scale or multi-unit projects, where repetition can lead to substantial productivity gains.

Material and Component Selection: Material and component selection is a vital element of buildability that directly affects construction methodology. Materials chosen for a project must not only meet structural, functional, and aesthetic requirements but also be suitable for efficient construction. Materials that are lightweight, durable, and easy to handle reduce labour effort, improve safety, and accelerate installation processes. Buildability considerations ensure that selected materials are compatible with available construction techniques and site conditions. For instance, materials that require specialised handling or installation methods may complicate construction methodology and increase costs. Conversely, materials that are readily available, easy to transport, and simple to install support efficient construction processes. Proper material selection also minimises waste and reduces the likelihood of damage during handling and installation, contributing to overall project efficiency and sustainability.

Prefabrication and Modular Construction: Prefabrication and modular construction are increasingly important elements of buildability in modern construction methodology. By shifting a significant portion of construction work from the site to controlled factory environments, prefabrication enhances quality control, reduces on-site labour requirements, and shortens construction timelines. Factory-based production allows components to be manufactured under standardised conditions, resulting in higher precision and consistency. From a buildability perspective, prefabrication simplifies on-site construction methodology by reducing the number of activities performed in constrained and variable site conditions. Modular construction enables rapid assembly of building elements, improving productivity and reducing weather-related delays. Additionally, prefabrication enhances safety by minimising hazardous on-site activities and reducing the need for working at heights. When integrated early in the design and planning stages, prefabrication and modularisation significantly improve construction efficiency and project outcomes.

Site Access and Logistics: Effective site access and logistics planning are essential elements of buildability in construction methodology. Buildable designs consider site constraints such as access routes, storage areas, lifting zones, and equipment movement from the outset. Proper logistics planning ensures a smooth flow of materials, equipment, and labour, minimising congestion and delays on-site. Construction methodology that incorporates buildability principles allows for efficient coordination of deliveries, storage, and installation activities. Clear access routes and designated storage areas reduce handling time and the risk of damage to materials. Poor site logistics, often resulting from inadequate buildability consideration, can lead to overcrowding, inefficient material handling, and increased safety risks. By addressing site access and logistics early, buildability enhances construction efficiency and reduces operational disruptions.

Safety and Risk Reduction: Safety and risk reduction are integral to buildability and construction methodology. Construction methodologies influenced by buildability principles aim to eliminate or minimise hazards through thoughtful design and planning. Simplifying construction tasks, reducing work at heights, and minimising heavy manual handling all contribute to safer working conditions. Buildable designs allow for safer construction methodologies by enabling the use of standard procedures, prefabricated components, and well-planned sequencing. Improved safety not only protects workers but also enhances productivity by reducing accidents, work stoppages, and disruptions. A safer construction environment fosters better morale and efficiency, contributing to successful project delivery.

In summary, the key elements of buildability in construction methodology, design simplicity and clarity, effective sequencing, standardisation, appropriate material selection, prefabrication, site logistics, and safety, collectively shape how construction projects are executed. By integrating these elements into construction methodology from the earliest stages, project teams can achieve higher efficiency, improved quality, reduced costs, and enhanced safety. Ultimately, a buildable construction methodology delivers greater value, supports sustainable practices, and ensures successful project outcomes throughout the construction lifecycle.

4.7 Chapter Conclusion

This chapter has comprehensively examined how buildability is embedded within procurement strategies, design development, construction practice, and long-term asset performance. It has been demonstrated that buildability is not a standalone technical consideration but a strategic project philosophy that must be integrated across the entire project lifecycle. By aligning procurement approaches, early stakeholder engagement, design decisions, and construction methodologies, buildability becomes a powerful enabler of efficiency, quality, safety, and sustainability.

The discussion on BLP highlighted the importance of structuring procurement systems to prioritise constructability from the outset. Traditional procurement methods often separate design and construction responsibilities, limiting opportunities for practical input during early decision-making. In contrast, BLP emphasises the early integration of construction expertise, flexible design evaluation, and performance-oriented selection criteria. By embedding buildability objectives into procurement frameworks, clients and project teams can reduce fragmentation, encourage innovation, and create collaborative environments that improve overall project outcomes.

ECI further reinforces the operationalisation of buildability principles. This chapter illustrated how ECI enables contractors to contribute their practical knowledge during critical design stages, enhancing sequencing strategies, risk identification, cost optimisation, and resource planning. When implemented effectively, ECI strengthens coordination, reduces rework, and improves certainty in project delivery. However, a successful application requires clear governance structures, trust-based relationships, and well-defined roles to overcome challenges such as contractual complexity and perceived cost implications. Best practice principles, including transparency, early engagement, and shared risk management, are essential to maximising the benefits of ECI.

Buildability in design and construction practice was presented as a continuous and iterative process. Design decisions directly influence construction methods, site logistics, material handling, and workforce productivity. Simplification, standardisation, modular coordination, and clear detailing enhance both efficiency and safety during execution. At the construction stage, effective planning, sequencing, and coordination translate buildable designs into practical reality. This chapter emphasised that buildability must be actively managed rather than assumed, requiring ongoing collaboration between designers, contractors, and clients.

The nexus between buildability and maintainability extends the concept beyond project completion. Designing for ease of maintenance, accessibility, durability, and whole-life performance ensures that assets remain functional, safe, and cost-effective throughout their lifecycle. This integration of maintainability considerations strengthens the long-term value proposition of buildability, reinforcing the importance of whole-life thinking in construction decision-making.

Finally, this chapter explored the conceptual link between buildability and construction methodology. Construction methods are not merely execution tools but strategic determinants of project performance. The selection of appropriate methodologies, considering sequencing, technology, workforce capability, and site conditions, must align with design intent and procurement strategy. When buildability principles inform methodological decisions, projects benefit from improved coordination, reduced waste, enhanced productivity, and minimised risk.

In conclusion, buildability in construction is a holistic framework that connects procurement, design, construction execution, and asset operation. Through BLP, ECI, integrated design practices, and methodical construction planning, buildability transforms project delivery from a fragmented process into a coordinated and value-driven system. Embedding these principles across all phases of construction ensures not only improved immediate project performance but also sustainable, long-term asset success.

Bibliography

Abkar, M.M.A., Yunus, R., Gamil, Y., & Albaom, M.A. (2024). Enhancing construction site performance through technology and management practices as material waste mitigation in the Malaysian construction industry. *Heliyon*, 10(7), e28721. https://doi.org/10.1016/j.heliyon.2024.e28721

Ahmed, F., Gkantou, M., Nikitas, G., Ferentinou, M., Bras, A., & Riley, M. (2025). A review of recent developments in structural elements of modular steel building systems. *Journal of Building Engineering*, 111, 113087. https://doi.org/10.1016/j.jobe.2025.113087

Aina, O.O. (2015). Buildability problems in construction projects in Nigeria: Causes and impacts. *European Scientific Journal*, 11(27), 353–366.

Akhmetzhanova, B., Nadeem, A., Hossain, M.A., & Kim, J.R. (2022). Clash detection using building information modeling (BIM) technology in the Republic of Kazakhstan. *Buildings*, 12(2), 102. https://doi.org/10.3390/buildings12020102

Bamisile, A. (2004). *Building production management*. Foresight Press Limited.

Blanchard, B.S., Verma, D., & Peterson, E.L. (1995). *Maintainability: A key to effective serviceability and maintenance management.* John Wiley & Sons. Inc.

Birolini, A. (2004). *Reliability engineering.* Springer-Verlag.

Borges, M.L.A.E., Granja, A.D., & Monteiro, A. (2024). A hybrid framework for multi-objective construction site layout optimization. *Buildings*, 14(12), 3790. https://doi.org/10.3390/buildings14123790

Botha, P.S., & Scheepbouwer, E. (2015). Relationship between early contractor involvement and financial performance in the rebuilding of infrastructure in Christchurch, New Zealand, *Transportation Research Record: Journal of the Transportation Research Board*, 2504, 66–72. https://doi.org/10.3141/2504-08

Building and Construction Authority. (2023). Design for maintenance. Available at: https://www1.bca.gov.sg/buildsg/facilities-management-fm/design-for-maintainability (accessed April 25, 2023).

Caterino, M., Cutolo, P., De Simone, V., Di Pasquale, V., Miranda, S., & Riemma, S. (2025). Workforce training strategies and performance assessment in manufacturing environments: A preliminary investigation. *6th international conference on Industry 4.0 and smart manufacturing, procedia computer science*, 253, 2399–2408. https://doi.org/10.1016/j.procs.2025.01.300

Chen, C. (2023). Application of modular construction method to constructing buildings: Advantages and barriers. In: *Proceedings of the institution of civil engineers - Smart infrastructure and construction*, 176(13), 1–9. https://doi.org/10.1680/jsmic.22.00017

Chew, M.Y.L. (2010). *Maintainability of facilities: For building professionals.* World Scientific Publishing Co. Pte. Ltd.

Construction Industry Institute (CII). (1995). *Pre – Project planning handbook.* University of Texas, Austin.

Cunningham, L.S., & Pomfret, M.A. (2007). Partnering contracts in practice at Blackpool, UK. Proceedings of the Institution of Civil Engineers. *Municipal Engineer*, 160(1), 17–21.

de Lima, D.M., Silva, W.W.N., López-Yánez, P.A., & Medeiros, I.D.S. (2024). Analysis and structural design of a building with multiple floors in composite steel and concrete structure. *Discover Civil Engineering*, 1, 150. https://doi.org/10.1007/s44290-024-00150-3

de Silva, N., Dulaimi, M.F., Ling, F.Y.Y., & Ofori, G. (2004). Improving the maintainability of buildings in Singapore. *Building and Environment*, 39(10), 1243–1251. https://dx. doi.org/10.1016/j.buildenv.2004.02.011

Dunston, P., & Williamson, C. (1999). Incorporating maintainability in constructability review process. *Journal of Management in Engineering*, 15(5), 56–60. https://doi.org/10.1061/(ASCE)0742-597X(1999)15:5(56)

Durdyev, S. (2021). Review of construction journals on causes of project cost overruns. *Engineering, Construction and Architectural Management*, 28(4), 1241–1260.

Eadie, R., & Matthew, G. (2014). Analysing the advantages of early contractor involvement. *International Journal of Procurement Management*, 7(6), 661–676. https://doi.org/10.1504/IJPM.2014.064971.

ELZomor, M., & Parrish, K. (2016). Investigating building construction process and developing a performance index. *International Conference on Sustainable Design, Engineering and Construction, Procedia Engineering*, 145(2016), 211–218. https://doi.org/10.1016/j.proeng.2016.04.063

Fadamiro, J.A., & Bobadoye, S. (2006). Managing the building design process for sustainability and improved quality. *Civil Engineering Dimension*, 8(1), 1–7.

Finnie, D., Ali, N.A., & Park, K. (2017). Enhancing off-site manufacturing through early contractor involvement (ECI) in New Zealand. In: *Proceedings of the institution of civil engineers, management, procurement and law,* 1700029. https://doi.org/10.1680/jmapl.17.00029.

Firoozi, A.A., Firoozi, A.A., & Maghami, M.R. (2025). Sustainable practices in geotechnical engineering: Forging pathways for resilient infrastructure. *Results in Engineering*, 26, 105577. https://doi.org/10.1016/j.rineng.2025.105577

Gorgolewski, M. (2008). Designing with reused building components: Some challenges. *Building Research & Information*, 36(2), 175–188. https://doi.org/10.1080/09613210701559499

Gransberg, D.D. (2013). Early contractor design involvement to expedite delivery of emergency highway projects: Case studies from six states. *Transportation Research Record: Journal of the Transportation Research Board*, 2347, 19–26. https://doi.org/10.3141/2347-03

Gransberg, D.D. (2016). Comparing construction manager–general contractor and federal early contractor involvement project delivery methods. *Transportation Research Record: Journal of the Transportation Research Board*, 2573, 18–25. https://doi.org/10.3141/2573-03

Haglund, P., Rudberg, M., & Sezer, A.A. (2022). Organizing logistics to achieve strategic fit in building contractors: a configurations approach. *Construction Management and Economics*, 40(9), 711–726. https://doi.org/10.1080/01446193.2022.2098990

Hastie, J., Sutrisna, M., & Egbu, C. (2017). Modelling knowledge integration process in early contractor involvement procurement at tender stage – A Western Australian case study. *Construction Innovation.* https://doi.org/10.1108/CI-04-2016-0021

Hostetler, M. (2010). Beyond design: The importance of construction and post-construction phases in green developments. *Sustainability*, 2010(2), 1128–1137. https://doi.org/10.3390/su2041128

Ibrahim, A., Zayed, T., & Lafhaj, Z. (2024). Enhancing construction performance: A critical review of performance measurement practices at the project level. *Buildings*, 14(7), 1988. https://doi.org/10.3390/buildings14071988

Jiang, L., & Leicht, R.M. (2016). Supporting automated constructability checking for formwork construction: An ontology. *Journal of Information Technology in Construction (ITcon)*, Special issue: CIB W78 2015 Special track on Compliance Checking, Vol. 21, pp. 456–478.

Jong, C.Y., Sim, A.K.S., & Lew, T.Y. (2019). The relationship between TQM and project performance: Empirical evidence from Malaysian construction industry. *Cogent Business & Management*, 6(1). https://doi.org/10.1080/23311975.2019.1568655

Lam, P.T.I., Wong, F.W.H., & Chan, A.P.C. (2006). Contributions of designers to improving buildability and constructability. *Design Studies*, 27(4), 457–479.

Lappalainen, E., Uusitalo, P., Pikas, E., Seppänen, O., Peltokorpi, A., Uusitalo, P., Reinbold, A., & Menzhinskii, N. (2022). Improving design quality by contractor involvement: An empirical study on effects. *Buildings*, 12(8), 1188. https://doi.org/10.3390/buildings12081188

Laryea, S., & Watermeyer R. (2016). Early contractor involvement in framework contracts. Proceedings of the institution of civil engineers. *Management, Procurement and Law*, 169(1), 4–16.

Lee, J., Park, M., Lee, H.S., Kim, T.Y., Kim, S., & Hyun, H. (2017). Workflow dependency approach for modular building construction manufacturing process using dependency structure matrix (DSM). *KSCE Journal of Civil Engineering*, 21(5), 1525–1535. https://doi.org/10.1007/s12205-016-1085-1

Li, S., Liu, L., & Peng, C. (2020). A review of performance-oriented architectural design and optimization in the context of sustainability: Dividends and challenges. *Sustainability*, 12(4), 1427. https://doi.org/10.3390/su12041427

Love, P.E.D., O'Donoghue, D., Davis, P.R., & Smith, J. (2014). Procurement of public sector facilities. *Facilities*, 32(9/10), 460–471. https://dx. doi.org/10.1108/F-03-2012-0020

Maelissa, N., Rohman, M., & Wiguna, I. (2023). Influencing factors of sustainable highway construction. *E3S Web of Conferences*, 429, 03002. https://doi.org/10.1051/e3sconf/202342903002

Mbamali, I., Aiyetan, O.A., & Kehinde, J.O. (2005). Building design for buildability: An investigation of the current practice in Nigeria. *Building and Environment*, 40(9), 1267–1274. https://doi.org/10.1016/j.buildenv.2004.10.006

Meng, X., & Humphreys, P. (2015). The application of early contractor involvement and its impact on project performance. *International Journal of Project Organisation and Management*, 7(4), 339–351.

Molenaar, K.R., Triplett, J.E., Porter, J.C., DeWitt, S.D., & Yakowenko, G. (2007). Early contractor involvement and target pricing in U.S. and UK highways. *Transportation Research Record: Journal of the Transportation Research Board*, 2040, 3–10. https://doi.org/10.3141/2040-01

Moses, T., Heesom, D., Oloke, D., & Crouch, M. (2016). Exploring the benefits of early contractor involvement (Eci) and the mechanism for automated 5D BIM quantification process in offsite modular construction, 43–50. https://doi.org/10.29173/mocs6

Mosey, D. (2009). *Early contractor involvement in building procurement: Contractors, partnering and project management*. Wiley Blackwell.

Mydin, S.H., Zin, R.M., Majid, M.Z.A., Zahidi, M., & Memon, A.H. (2011). Buildability attributes at design phase in Malaysian building construction. *International Journal of Sustainable Construction Engineering & Technology*, 2(1), 24–43.

Nibbelink, J.G., Sutrisna, M., & Zaman, A.U. (2017). Unlocking the potential of early contractor involvement in reducing design risks in commercial building refurbishment projects – A Western Australian perspective. *Architectural Engineering and Design Management*. https://doi.org/10.1080/17452007.2017.1348334

Norton, J.N., & Brouwer, A.B. (2009). The planning, design and construction process. In: Hessler, J.R., & Lehner, N.D.M. (Eds.) *American college of laboratory animal medicine, planning and designing research animal facilities* (pp. 17–44). Academic Press. https://doi.org/10.1016/B978-0-12-369517-8.00004-9

Osuizugbo, I.C. (2020). The need for and benefit of buildability analysis: Nigeria as a case study. *Journal of Engineering, Design and Technology*, 19(5), 1207–1230. https://doi.org/10.1108/JEDT-08-2020-0338

Osuizugbo, I.C. (2025a). Buildability-led procurement: A new model for future construction projects delivery. *Construction Innovation: Information, Process, Management*. https://doi.org/10.1108/CI-10-2023-0272

Osuizugbo, I.C. (2025b). Measures for improving the buildability of building designs in construction industry. *Construction Innovation: Information Process Management*, 25(2), 490–509. https://doi.org/10.1108/CI-08-2022-0223

Osuizugbo, I.C., Fauzi, M.A., Omer, M., Njogo, B.O., & Orekan, A.A. (2024a). Research trends on early contractor involvement in construction projects: A bibliometric analysis. *Construction Economics and Building*, 24(4/5), 160–178. https://doi.org/10.5130/AJCEB.v24i4/5.9113

Osuizugbo, I.C., Okolie, K.C., Oshodi, O.S., & Oyeyipo, O.O. (2023). Buildability in the construction industry: A systematic review. *Construction Innovation: Information Process Management*, 23(5), 1300–1322. https://doi.org/10.1108/CI-05-2022-0112

Osuizugbo, I.C., Omer, M.M., Rahman, R.A., & Oshodi, O.S. (2024b). A systematic search and review of early contractor involvement in construction project development. *Built Environment Project and Asset Management*, 14(6), 835–857. https://doi.org/10.1108/BEPAM-01-2024-0012

Owolabi, T.O.S., Harry, E.G., Adewumi, B.J., Onamade, A.O., & Alagbe, O.A. (2024c). Ensuring quality in construction project: The role of specifications as quality assurance tools. *Anchor University Journal of Science and Technology*, 5(2), 181–191.

Pheng, L.S., Gao, S., & Lin, J.L. (2015). Converging early contractor involvement (ECI) and lean construction practices for productivity enhancement: Some preliminary findings from Singapore. *International Journal of Productivity and Performance Management*, 64(6), 831–852. https://doi.org/10.1108/IJPPM-02-2014-0018

Rahman, M., & Alhassan, A. (2012). A contractor's perception on early contractor involvement. *Built Environment Project and Asset Management*, 2(2), 217–233. https://doi.org/10.1108/20441241211280855

Rahmani, F. (2020). Challenges and opportunities in adopting early contractor involvement (ECI): Client's perception. *Architectural Engineering and Design Management*. https://doi.org/10.1080/17452007.2020.1811079

Rahmani, F., Khalfan, M.M.A., & Maqsood, T. (2016). Analysing the drivers for early contractor involvement adoption by construction clients. *International Journal of Procurement Management*, 9(4), 373–397.

Rahmani, F., Khalfan, M.M.A., & Maqsood, T. (2018). A comparative study of early contractor involvement and project alliancing. *International Journal of Project Organisation and Management*, 10(2), 93–108.

Retterer, B.L., & Kowalski, R.A. (1984). Maintainability: A historical perspective. *IEEE Transactions on Reliability*, R-33(1), 56–61. https://doi.org/10.1109/TR.1984.6448275

Ruzieh, A.S., Awwad, B.S., & Razia, B.S. (2025). Review of logistics challenges within the construction industry. *EuroMid Journal of Business and Tech-Innovation*, 4(1), 1–22. https://doi.org/10.51325/ejbti.v4i1.206

Sagvekar, S., & Wayal, A.S. (2019). Early contractor involvement (ECI): Indian scenario of construction project delivery. *International Journal of Scientific & Technology Research*, 8(8), 807–811.

Scheepbouwer, E., & Humphries, A.B. (2011). Transition in adopting project delivery method with early contractor involvement. *Transportation Research Record: Journal of the Transportation Research Board*, 2228, 44–50. https://doi.org/10.3141/2228-06

Serpell, A., & Alarcón, L.F. (1998). Construction process improvement methodology for construction projects. *International Journal of Project Management*, 16(4), 215–221. https://doi.org/10.1016/S0263-7863(97)00052-5.

Sharma, S.N., Dehalwar, K., Singh, J., & Kumar, G. (2025). Prefabrication building construction: A thematic analysis approach. In: Singh, S.B., Gopalarathnam, M., & Roy, N. (Eds.) *Proceedings of the 3rd international conference on advances in concrete, structural, and geotechnical engineering—Volume 2*. ACSGE 2024. Springer Proceedings in Materials, vol 30. Springer. https://doi.org/10.1007/978-981-96-0751-8_28

Sholanke, A.B., Adeboye, A.B., & Alagbe, A.O. (2019). Design solutions creating barriers to achieving universal design compliance of academic buildings in universities in Nigeria. *International Journal of Civil Engineering and Technology*, 10(01), 671–690.

Song, L., Mohamed, Y., & Abourizk, S.M. (2009). Early contractor involvement in design and its impact on construction schedule performance. *Journal of Management in Engineering*, 25(1), 12–20.

Turner, N., & Riding, M. (2015). Early contractor involvement in Australia: Learnings from transfield services projects. *Small Enterprise Research*, 22(2–3), 173–184. https://doi.org/10.1080/13215906.2015.1061724

van der Walt, J.D., Botha, P.S., & Scheepbouwer, E. (2019). When to engage the contractor for pre-construction services in New Zealand. *Transportation Research Record*, 2673(2), 562–570. https://doi.org/10.1177/0361198118823516

Wondimu, P., Klakegg, O., & Lædre, O. (2020). Early contractor involvement (ECI): Ways to do it in public projects. *Journal of Public Procurement* (ahead-of-print). https://doi.org/10.1108/JOPP-03-2019-0015

Zakaria, Z., Ismail, S., & Md Yusof, A. (2013). Buildability and maintainability in final account closing: The greatest challenge for construction industry. *Conference Paper*, March 2013. https://doi.org/10.13140/RG.2.1.1758.2806

Zhang, K., & Tsai, J.-S. (2021). Identification of critical factors influencing prefabricated construction quality and their mutual relationship. *Sustainability*, 13(19), 11081. https://doi.org/10.3390/su131911081

Zhang, W., & Xu, J. (2022). Advanced lightweight materials for Automobiles: A review. *Materials & Design*, 221, 110994. https://doi.org/10.1016/j.matdes.2022.110994

5 Buildability Assessment in Practice

5.1 Buildability Assessment Concept

Buildability assessment is a critical and systematic process aimed at improving the ease, efficiency, and effectiveness with which a building can be constructed. It serves as an important tool for enhancing buildability by focusing on the thorough examination of building production information before construction activities commence on-site. The fundamental purpose of buildability assessment is to ensure that design solutions are not only aesthetically pleasing and functionally adequate but also practical, economical, and achievable under real construction conditions.

Buildability assessment can be defined as the critical examination of building production information with the objective of identifying, evaluating, and eliminating design-related issues that may lead to waste, inefficiency, rework, delays, or construction difficulties. This process involves preparing a detailed report, discussing the findings with designers and other project stakeholders, and ensuring that all identified impediments to construction are addressed at the design stage. By doing so, buildability assessment helps remove unnecessary complexity, wasted effort, and potential defects from the design before work begins on-site.

One of the key functions of buildability assessment is achieved through modifying design characteristics and revising specifications that contribute to constructability problems. This may include changes to element sizes, site layout plans, reinforcement detailing, dimensional tolerances, or material specifications. For example, excessive reinforcement congestion may be redesigned to allow easier placement and compaction of concrete, or overly tight dimensional tolerances may be adjusted to reflect realistic construction capabilities. These amendments help align design intent with construction practicality, thereby reducing errors and improving overall project performance.

Buildability assessment is primarily a physical and production-oriented evaluation of design information. Unlike purely conceptual design reviews, it focuses on how the design will be built in practice. As such, it plays a significant role in both the design and construction stages of building projects. The assessment involves a comprehensive examination of building production information, which typically includes drawings, specifications, schedules, and construction details. This examination entails several key activities.

DOI: 10.1201/9781003747284-7

First, it involves carefully reviewing and closely inspecting project documents in relation to their production requirements. This ensures that drawings and specifications are complete, coordinated, and free from ambiguities that could confuse contractors on-site. Second, the process includes inquiring into the design and specifications to understand the rationale behind certain design decisions and to assess whether alternative solutions could improve constructability. Third, it requires investigating and questioning the use of new, unfamiliar, or questionable materials, components, or construction details, especially where limited construction experience exists. This helps prevent the adoption of solutions that may be difficult to implement or maintain.

Additionally, buildability assessment draws upon knowledge and lessons learned from previous projects. Testing the design against past experience allows potential problems to be identified early and proven solutions to be applied. Finally, the assessment evaluates whether the design and specifications are capable of performing their intended functions while meeting durability, safety, and quality requirements. This holistic approach ensures that the final design is not only buildable but also fit for purpose over the building's lifecycle.

The absence of buildability assessment represents a significant gap in the evaluation of production information. Without it, designs may proceed to construction without being tested for production-friendliness, suitability, or freedom from defects. This often results in costly design changes, delays, disputes, and compromised quality during construction. Therefore, buildability assessment is essential for bridging the gap between design intent and construction reality.

For buildability assessment to be effective, it should commence simultaneously with initial project planning and early design stages. Early involvement allows potential construction challenges to be addressed proactively rather than reactively. The assessment focuses on buildability attributes that contribute to efficient construction, such as simplicity of design details, ease of material handling, and straightforward construction methods. Clear communication through well-coordinated drawings and specifications is also a critical consideration.

Other important factors considered during buildability assessment include standardisation and rationalisation of building components, repetition of elements, and the minimisation of unnecessary variations. These measures improve efficiency, reduce errors, and enhance productivity on-site. Site constraints and characteristics, such as access, space limitations, ground conditions, and the surrounding environment, are also evaluated to ensure that the design responds appropriately to actual site conditions. Furthermore, challenges commonly faced by construction teams during execution, such as sequencing difficulties or safety concerns, are considered to achieve better design outcomes.

Buildability assessment is traditionally one of the core responsibilities of professional builders, whose primary objective is to deliver value to clients through cost optimisation, time efficiency, and achieving quality outcomes at the first attempt. However, despite its benefits, there has often been resistance to builder-led input during the design process. Architects, who traditionally lead the construction process, may perceive buildability guidance as intrusive, unoriginal, or unnecessary, believing that such considerations have already been addressed during design development. This perception can limit effective collaboration and reduce the potential benefits of buildability assessment.

Nevertheless, the successful implementation of buildability assessment delivers significant benefits. These include a reduction in buildability-related problems, improvements in construction productivity, contributions to industry and economic development, and the promotion of buildability as a standard practice within the construction industry. Ultimately, buildability assessment fosters better integration between design and construction, leading to higher-quality projects delivered more efficiently and sustainably.

5.2 The Need for Buildability Assessment

Past experience within the construction industry has highlighted numerous challenges associated with achieving improved buildability in building projects. These challenges have underscored the necessity for a formal and systematic buildability assessment process. Despite advancements in design tools and construction technologies, many projects continue to suffer from inefficiencies, delays, cost overruns, and quality issues that can be traced back to inadequate consideration of buildability during the design stage. As a result, buildability assessment has emerged as an essential mechanism for addressing these persistent problems.

One of the major issues necessitating buildability assessment is the tendency of design professionals to operate independently. Architects, structural engineers, and services engineers often make design decisions in isolation, without sufficiently considering how these decisions affect other disciplines or the eventual construction process. This silo-based approach creates professional "walls" that hinder collaboration and integration. Consequently, design solutions that may be technically sound within a single discipline can become problematic when viewed from a construction perspective. Buildability assessment helps break down these barriers by encouraging interdisciplinary collaboration and early evaluation of design decisions from a construction standpoint.

Another challenge relates to concerns over the capability of architects to manage and supervise construction projects effectively. While architects are traditionally the leaders of the construction process, their training and focus are often design-oriented rather than production-oriented. Experience has shown that better and more practical design solutions are achieved when constructors are actively involved in the design process. Constructors bring valuable knowledge of construction methods, sequencing, labour productivity, material availability, and site constraints. Buildability assessment provides a formal platform for incorporating this construction expertise into the design process, leading to more practical and efficient outcomes.

Construction clients have also expressed increasing dissatisfaction with poor project delivery, particularly where buildability problems contribute to extended project durations and escalating costs. Delays caused by design changes, rework, and construction difficulties not only increase project costs but also undermine client confidence in the construction industry. Studies have revealed that many of these issues stem from the inability of designers and constructors to develop a common understanding of the entire construction process. The fragmented nature of the construction industry and the weak correlation between design and construction further exacerbate these problems. Buildability assessment directly addresses these shortcomings by promoting integration, coordination, and shared understanding among all project participants.

The need for buildability assessment can be further understood through several interrelated factors. First, there is a general lack of unity between designers and constructors. This lack of unity often results in adversarial relationships rather than collaborative partnerships. Buildability assessment encourages teamwork by aligning design decisions with construction realities. Second, designers and constructors frequently lack a shared understanding of the full construction process, leading to unrealistic designs and inefficient construction methods. Through systematic assessment, design solutions are evaluated against practical construction considerations.

High building costs also highlight the need for buildability assessment. Poor buildability often results in inefficient use of labour, materials, and equipment, leading to increased costs. In many cases, designs are not realised in the most economical way, preventing clients from attaining optimum returns on their investment. Buildability assessment helps optimise design solutions to achieve cost-effective construction without compromising quality or performance.

Site-related challenges further reinforce the importance of buildability assessment. Many construction sites are jam-packed with limited space, restricted access, and surrounding environmental or infrastructural constraints. Without proper assessment, designs may fail to accommodate these conditions, leading to logistical difficulties and safety risks. This is particularly critical for complicated projects involving complex forms, advanced technologies, or intricate building services systems that require careful coordination and harmonisation.

Defective or poorly prepared working drawings are another major contributor to buildability problems. Incomplete, ambiguous, or inconsistent drawings can result in errors, rework, and disputes during construction. Buildability assessment allows such defects to be identified and corrected early, improving clarity and reducing uncertainty on-site. Similarly, substandard design quality and poor correlation between design and construction necessitate systematic assessment to ensure that designs are both technically sound and constructible.

The lack of construction experience and practical knowledge among some designers further underscores the need for buildability assessment. When designs are developed without meaningful input from constructors, they may overlook practical considerations such as build sequence, material handling, and tolerances. This issue is compounded when contractors are not adequately informed about proposed design changes, leading to confusion and inefficiencies during execution. Buildability assessment promotes transparency and ensures that all parties are aware of and aligned with design intentions.

Communication failures represent another critical factor driving the need for buildability assessment. Ineffective communication among project stakeholders often leads to misunderstandings, delays, and conflicts. This problem is exacerbated when construction input is requested too late in the project lifecycle, at a point where design changes are costly and disruptive. Buildability assessment encourages early and continuous communication, enabling issues to be addressed proactively.

Financial constraints also play a role. Clients and project teams are often unwilling to invest additional time, money, and effort during the early stages of a project, despite the long-term benefits. This reluctance, combined with ignorance of the buildability concept, limits the adoption of best practices. Moreover, the lack of competent personnel to carry out buildability improvements further hinders effective implementation.

Finally, the fragmented nature of the construction industry remains a fundamental reason for the need for buildability assessment. Fragmentation leads to poor coordination, misaligned objectives, and inefficiencies throughout the project lifecycle. Buildability assessment serves as a unifying process that integrates design and construction, fosters collaboration, and promotes shared responsibility for project success.

In summary, the need for buildability assessment arises from a combination of technical, organisational, economic, and cultural challenges within the construction industry. By addressing issues such as poor integration, high costs, site constraints, inadequate communication, and lack of construction input during design, buildability assessment plays a vital role in improving project performance and delivering better value to clients.

5.3 The Benefits of Buildability Assessment

The practice of buildability assessment offers numerous benefits to the construction industry and its stakeholders. When the concept of buildability assessment is well understood and effectively implemented, it provides significant advantages to clients, designers, and constructors alike. Consequently, integrating buildability assessment into the construction process is widely recognised as a valuable approach for improving project performance and achieving sustainable development within the construction sector. The benefits of buildability assessment may be categorised into direct and indirect benefits, both of which contribute to enhanced project outcomes.

5.3.1 Direct Benefits of Buildability Assessment

One of the most significant direct benefits of buildability assessment is that it makes construction planning easier and more efficient. By identifying potential construction difficulties during the design stage, project teams can develop realistic construction methods, sequences, and resource plans. This proactive approach reduces uncertainty and minimises disruptions during construction.

Buildability assessment also contributes to the reduction of design and construction costs. Early identification of inefficient design details, complex construction methods, and impractical specifications allows for cost-effective alternatives to be explored. These design optimisations help eliminate unnecessary expenses associated with rework, delays, and material wastage. As a result, overall project costs can be significantly reduced.

Another major benefit is the shortening of construction schedules. Designs that are assessed for buildability are generally simpler, more standardised, and better coordinated, allowing construction activities to proceed more smoothly and efficiently. Reduced construction complexity and improved sequencing led to faster project completion and better utilisation of resources.

Quality improvement is another direct outcome of buildability assessment. When construction processes are carefully considered during design, the likelihood of defects, errors, and poor workmanship is reduced. Clearer drawings, practical details, and achievable tolerances contribute to higher construction quality and improved performance of the completed building.

Buildability assessment also promotes realistic responsibilities and commitments among project participants. By involving constructors early in the design process, roles and expectations are clearly defined, reducing conflicts and misunderstandings during construction. Additionally, the owner's role in the project begins as early as possible, enabling clients to make informed decisions regarding cost, time, and quality from the outset.

5.3.2 Indirect Benefits of Buildability Assessment

In addition to its direct benefits, buildability assessment delivers numerous indirect benefits that enhance collaboration and innovation within the construction industry. One such benefit is the promotion of teamwork among stakeholders. Buildability assessment encourages designers, constructors, and clients to work together toward a shared vision and common project objectives, fostering a collaborative project environment.

Through this collaborative approach, each stakeholder works toward mutual benefit rather than pursuing isolated interests. The assessment process facilitates the crossing and integration of scientific and technical disciplines, enabling architects, engineers, and constructors to share knowledge and perspectives. This interdisciplinary collaboration enhances problem-solving and leads to more holistic design solutions.

Experience transfer is another key indirect benefit. Buildability assessment provides a platform for constructors to share practical construction experience with designers, while designers can communicate design intent more effectively. As a result, contractors gain a better understanding of design objectives, and designers develop a deeper appreciation of construction challenges and site realities.

Buildability assessment also opens opportunities for innovation in both design and construction. By encouraging early dialogue and joint problem-solving, innovative construction methods, materials, and technologies can be explored. This collaborative environment supports the development and adoption of labour-saving and cost-effective construction techniques.

Another indirect benefit is the shortening of the learning curve for project teams. Repeated application of buildability assessment enhances collective knowledge and competence, allowing teams to learn from past experiences and apply lessons more effectively in future projects. Over time, this accumulated knowledge becomes a competitive advantage, enabling organisations to improve performance and compete more effectively in the construction business.

5.3.3 Additional Benefits of Buildability Assessment

Numerous studies have demonstrated that buildability assessment is highly beneficial to the construction industry. One key benefit is its ability to identify cost-effective methods of site construction and future maintenance strategies. Buildability assessment also helps draw designers' attention to aspects of working drawings that may hinder the achievement of client objectives, allowing such issues to be resolved early.

Timely and proper consideration of buildability ensures that the final design meets established performance criteria while achieving cost savings. This includes reducing material waste, adopting labour-saving technologies, and improving construction efficiency. Increased efficiency, elimination of wasteful efforts, and improved quality are among the most notable outcomes of effective buildability assessment.

Further benefits include cost advantages for designers, constructors, and clients through reduced project costs and improved profit margins. Projects are more likely to be completed ahead of schedule or within planned timeframes, enhancing client satisfaction and reducing financial risks. Quality improvements contribute to smoother project delivery and reduced post-construction defects.

Buildability assessment also supports the principle of "getting it right at the first and only attempt." By resolving potential problems during design, the need for variations, redesign, and rework during construction is minimised. This not only saves time and money but also reduces the administrative burden associated with contract variations and disputes, particularly for architects and project managers.

The adoption of labour-saving construction technologies and methods further enhances productivity and reduces reliance on skilled labour, which is often scarce. Additionally, minimisation of material waste contributes to sustainability objectives and environmental responsibility.

Ultimately, the cumulative effect of these benefits is increased profitability for all stakeholders and improved overall performance of the construction industry. By integrating buildability assessment into standard practice, projects can be delivered more efficiently, economically, and with higher quality, thereby reinforcing its importance as a vital component of modern construction management.

5.4 Information Requirements of Buildability Assessment

Effective buildability assessment depends largely on the availability and accuracy of relevant project information. The quality of the assessment is directly influenced by how comprehensive, up-to-date, and coordinated the information is. The information required for buildability assessment can be broadly classified into two main categories: project production information and regulatory information. Both categories are essential to ensure that design solutions are practical, compliant, and achievable under real construction conditions.

5.4.1 Project Production Information

Project production information forms the core input for buildability assessment, as it defines the project objectives, design intent, and construction requirements. The client's brief is a fundamental document, as it outlines the client's needs, functional requirements, quality expectations, budget constraints, and project timelines. Understanding the brief ensures that buildability considerations align with the client's objectives.

The survey plan and site analysis report provide critical information on-site boundaries, topography, access, existing utilities, and surrounding conditions. These documents help assess site constraints that may affect construction methods, logistics, and sequencing. Similarly, the environmental impact assessment (EIA) report identifies environmental sensitivities, restrictions, and mitigation measures that influence construction planning and material selection.

The site investigation and soil test report are vital for assessing ground conditions and determining appropriate foundation systems and construction techniques. Poor understanding of soil conditions can lead to impractical designs and costly changes during construction.

Design information in the form of architectural, structural, and building services drawings, schedules, and specifications is central to buildability assessment. These documents are examined for clarity, coordination, and constructability. Architectural information defines spatial layouts and finishes, structural information addresses load-bearing systems and detailing, while services information ensures effective integration of mechanical, electrical, and plumbing systems. Specialist drawings, such as acoustic or fire engineering details, further support the assessment of complex or performance-driven design elements.

Finally, the preliminary estimate or client's budget provides a financial framework within which buildability decisions must be made. Cost information allows assessors to evaluate whether proposed designs and construction methods are economical and aligned with budgetary limits.

5.4.2 Regulations

Regulatory information ensures that the project complies with statutory and legal requirements. Physical development regulations guide land use, setbacks, height restrictions, and density controls, which directly affect design feasibility. Building control regulations and building byelaws establish minimum standards for safety, health, accessibility, and structural integrity. Environmental regulations govern issues such as waste management, emissions, noise, and sustainability requirements.

In conclusion, for buildability assessment to be effective, current and complete copies of all relevant project production and regulatory documents must be readily available. Access to accurate information enables informed evaluation, supports practical design decisions, and ensures that construction challenges are addressed early in the project lifecycle.

5.5 Buildability Assessment Implementation in Practice

Buildability assessment in practice is a structured and planned procedure aimed at improving the quality, clarity, and effectiveness of building production information. It translates the theoretical concept of buildability into actionable steps that can be applied throughout the design stage of a construction project. As earlier established, the primary function of buildability assessment is to critically examine production information with the aim of identifying and evaluating design-related issues, preparing a comprehensive report, discussing the findings with designers and other stakeholders, and ensuring the removal of waste, wasted effort, and impediments to construction before work commences on-site.

In practical terms, buildability assessment is not a one-off activity but an iterative process that evolves alongside design development. Figure 5.1 illustrates a framework

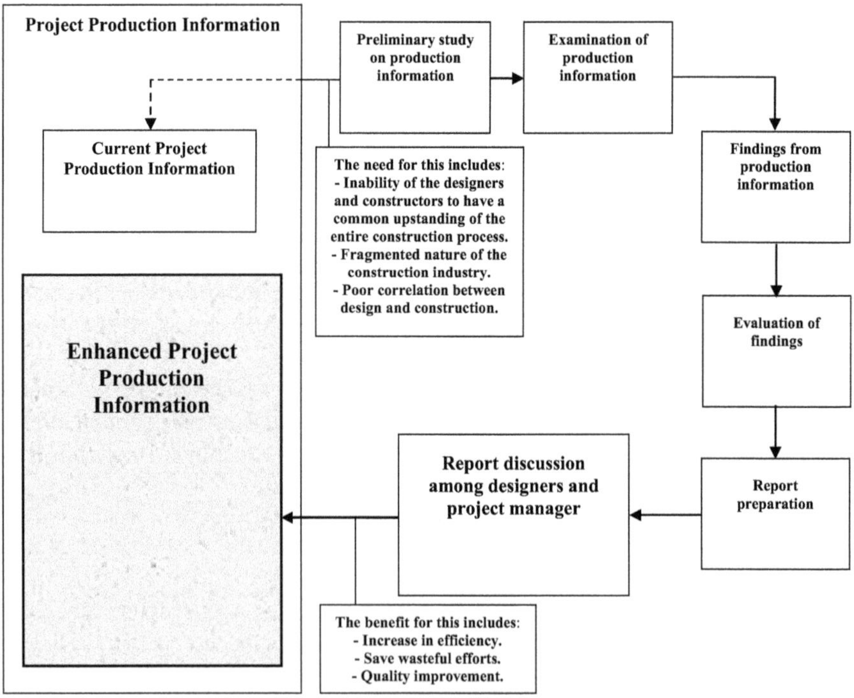

Figure 5.1 Framework for buildability assessment on project production information.

Source: Osuizugbo (2020).

for conducting buildability assessment on project production information, highlighting the sequential and logical steps involved in achieving production-friendly design solutions.

The first step in carrying out buildability assessment is a preliminary study of the available production information. This initial review involves familiarisation with key project documents such as the client's brief, survey plan, site analysis report, architectural drawings, structural drawings, building services drawings, specifications, and other relevant reports. At this stage, particular attention is paid to site-related issues, including access, topography, surrounding developments, environmental constraints, and existing infrastructure. Understanding the site context early enables potential construction challenges to be anticipated and addressed proactively.

Following the preliminary study, the next step involves a detailed and systematic examination of the production information. This requires careful and close inspection of working documents in relation to how the building will be produced on-site. The examination aims to identify omissions, errors, inconsistencies, and discrepancies within and between different sets of drawings and specifications. It also focuses on assessing the adequacy of design details, the compatibility of architectural, structural, and services information, and the practicality of proposed construction methods. Issues such as complex detailing, unrealistic tolerances, reinforcement congestion, difficult material handling, and unclear sequencing are typically uncovered during this stage.

Once these issues have been identified, the next phase involves evaluation. The findings from the examination are assessed against key project parameters, including the client's brief, preliminary cost estimate or budget, and applicable regulations. This evaluation determines whether the identified issues compromise project objectives in terms of cost, time, quality, safety, and performance. The aim is not to criticise design intent but to assess whether alternative solutions could improve constructability while still meeting functional, aesthetic, structural, and regulatory requirements.

Based on the evaluation, a buildability assessment report is then prepared. This report clearly documents the identified problems, their potential impacts on construction, and recommended solutions or design modifications. The emphasis of the report is on eliminating waste, reducing wasted effort, and removing obstructions to site execution from the production information. Importantly, recommended changes must not compromise the aesthetics, functionality, structural stability, durability, or serviceability of the building. Instead, the focus is on making the design more production-friendly, efficient, and economical.

The final step in buildability assessment in practice is the discussion and review of the assessment report. A formal discussion forum is held involving key project stakeholders, including builders, designers, and the project manager. This collaborative session allows findings and recommendations to be explained, debated, and refined. It also provides an opportunity for designers to clarify design intent and for builders to share construction expertise. Through constructive dialogue, agreed modifications are incorporated into the design, ensuring that all parties have a shared understanding of the final production information.

Once this process is completed, enhanced project production information emerges. The revised drawings and specifications are clearer, more coordinated, and better aligned with construction realities. As a result, the project is better positioned for efficient execution on-site, with reduced risk of delays, rework, and disputes. In this way, buildability assessment in practice serves as a vital link between design and construction, ensuring that projects are delivered more effectively and successfully.

5.6 Drivers for Buildability Assessment Implementation in the Construction Industry

The implementation of buildability assessment in the construction industry is influenced by a wide range of technical, organisational, cultural, and institutional drivers. Understanding these drivers is essential, as it enables construction stakeholders to develop appropriate strategies that encourage the effective adoption and sustained practice of buildability assessment. When these drivers are present and actively supported, buildability assessment can be successfully embedded within the project delivery process, leading to improved project outcomes and industry performance.

One of the most significant drivers of buildability assessment implementation is the owner's and client's commitment. Clients who recognise the long-term benefits of buildability assessment, such as reduced costs, improved quality, and timely project delivery, are more willing to invest additional time, effort, and resources during the early stages of a project. Clients' awareness of the value of buildability assessment on building design strongly influences project teams to prioritise constructability considerations and collaborative decision-making.

Another key driver is the extent to which designers value and prioritise buildability in their designs. When designers consider buildability as an integral design objective rather than an afterthought, they are more likely to produce designs that are practical and efficient to construct. Designers' adequate knowledge of construction operations further enhances this process, as it allows them to anticipate site challenges and develop solutions that align with construction realities. Education and training on buildability, therefore, play a critical role in strengthening designers' competence and confidence in addressing buildability-related issues.

The completeness of design information before construction commences is also a major driver. When designs are sufficiently developed and coordinated at the start of construction, the need for frequent changes and rework is significantly reduced. Closely related to this is the need for adequate coordination among different design disciplines, including architectural, structural, and building services teams. Effective coordination ensures that design requirements on aesthetics and functionality do not contradict buildability considerations and that conflicts between disciplines are resolved early.

An effective communication and coordination channel between designers and constructors at the design stage is another critical driver. Early and continuous interaction allows constructors to contribute practical insights on materials, methods, sequencing, and logistics. The early involvement of contractors helps designers develop more buildable designs and ensures that proposed changes are properly reviewed and addressed. This collaborative approach fosters mutual understanding and reduces adversarial relationships.

The ability of designers to solve buildability problems and their availability of adequate time to focus on buildability issues further support implementation. Designers who are pressured by tight schedules may overlook buildability considerations, whereas sufficient time allocation encourages thoughtful evaluation and refinement of design solutions. Incentives for designers to develop more buildable designs also serve as a motivating factor, reinforcing positive behaviour and innovation.

Procurement approaches significantly influence buildability assessment implementation. The minimisation of traditional procurement methods and the adoption of integrated approaches, such as design-and-build procurement systems, encourage collaboration and shared responsibility among project participants. Integrated procurement systems promote team spirit, unity among professionals, and collective ownership of project outcomes, all of which are essential drivers of buildability assessment.

Government involvement and institutional support also play an important role. Adequate support from the government, through policies, regulations, guidelines, and incentives, can encourage wider adoption of buildability assessment practices. Public sector leadership in mandating or promoting buildability assessment on projects sets a strong example for the private sector and helps institutionalise the practice across the industry.

Other organisational drivers include management commitment and supportive policies within construction firms. When organisational leadership prioritises buildability assessment, provides resources, and establishes clear procedures, implementation becomes more consistent and effective. Receptiveness of consultants and openness to collaborative input further enhance the success of buildability assessment.

Technological and contextual factors also act as drivers. The technological expertise of contractors, availability of local resources, and effective project scheduling influence how buildability considerations are integrated into design and construction planning. Advanced construction technologies and skilled contractors enable innovative and efficient solutions, while realistic scheduling allows adequate time for assessment and coordination.

Finally, intangible but equally important drivers include team spirit, unity among professionals, and awareness of buildability assessment across the industry. A collaborative culture encourages knowledge sharing, experience transfer, and collective problem-solving. When stakeholders share a common vision and understanding of the benefits of buildability assessment, implementation becomes more natural and sustainable.

In summary, the drivers for buildability assessment implementation are largely within the control of construction stakeholders, including clients, designers,

Table 5.1 Drivers for buildability assessment implementation

S/No	*Drivers*
1	Designers value buildability in their designs
2	Designers' adequate knowledge of construction operations
3	Designs are complete when construction starts
4	Adequate co-ordination amongst different design disciplines
5	Design requirements on aesthetics do not contradict buildability considerations
6	Designers are capable to solve buildability problems
7	Designers have adequate time to give sufficient attention to buildability
8	Adequate channel for co-ordination and communication between designers and constructors at the design stage
9	Designers develop buildable designs
10	Proper attention to proposed changes by contractors
11	Designers know the benefits of improved buildability of their designs
12	Education and training on buildability are provided to designers
13	Incentives are provided to designers for developing more buildable designs
14	Designers consider buildability important
15	Minimisation of traditional procurement approach
16	Design requirements on functionalities do not contradict buildability considerations
17	Adequate support from the government
18	Willingness to invest extra money and effort during the early stage of project
19	Awareness of buildability assessment
20	Unity amongst different professionals in the construction industry
21	Team spirit
22	Clients' awareness of the benefits of conducting buildability assessment on building design
23	Involvement of contractors early in the design process
24	Integrated approach/design and build procurement system
25	Proper scheduling
26	Technological expertise of the contractor
27	Owner's commitment
28	Receptiveness of the consultants
29	Local resource availability
30	Management's commitment and policies

Source: Osuizugbo et al. (2022).

contractors, consultants, management, and government bodies. As highlighted in Table 5.1, factors such as commitment, awareness, coordination, education, procurement strategies, and supportive policies collectively determine the success of buildability assessment implementation. By recognising and strengthening these drivers, the construction industry can effectively embed buildability assessment as a standard practice, leading to improved efficiency, quality, and value in project delivery.

5.7 Barriers to Buildability Assessment Implementation in the Construction Industry

Despite the numerous benefits associated with buildability assessment, its implementation in the construction industry remains limited due to several persistent barriers. These barriers hinder the effective integration of buildability considerations into the design and construction process, thereby reducing the potential gains in efficiency, cost savings, and quality improvement. Understanding these barriers is essential for developing practical strategies to enhance the implementation of buildability assessment. Importantly, most of these barriers are within the influence and control of key construction stakeholders, including clients, designers, contractors, professional bodies, and government agencies.

One of the major barriers to buildability assessment implementation is poor coordination among different design disciplines. Architectural, structural, and building services designers often work in silos, resulting in fragmented design outputs that are poorly integrated. This lack of coordination leads to clashes, inconsistencies, and impractical design solutions that negatively affect constructability. When coordination is weak, buildability assessment becomes difficult to conduct effectively, as resolving conflicts requires significant redesign effort.

Another significant barrier is the lack of awareness of buildability assessment within the industry. Many stakeholders, particularly designers and clients, are either unfamiliar with the concept or do not fully understand its benefits. This lack of awareness limits motivation to adopt buildability assessment practices and undermines efforts to institutionalise them. Closely related to this is the absence of education and training on buildability, which leaves designers ill-equipped to incorporate construction considerations into their designs.

The design review pattern commonly adopted in many projects also presents a barrier. In this pattern, construction personnel or builders are invited to review building production information only after major design decisions have been made. At this late stage, meaningful buildability input is often constrained, and proposed changes may be resisted or ignored. This reactive approach reduces the effectiveness of buildability assessment and reinforces the separation between design and construction.

Financial and organisational constraints further impede implementation. Many clients and project teams show an unwillingness to invest extra time, money, and effort during the early stages of a project, even though early investment yields long-term benefits. Tight project schedules also mean that designers often lack sufficient time to give adequate attention to buildability issues. As a result, buildability considerations are either overlooked or deferred until construction begins.

A strong cultural emphasis on aesthetics and functionality over buildability is another major barrier. Designers may prioritise visual appeal and functional performance without fully considering construction practicality. In many cases, design requirements related to aesthetics and functionality directly contradict buildability considerations, leading to complex details, unusual forms, and inefficient construction methods. This challenge is compounded when designers do not regard buildability as an important design criterion or are not required contractually to develop buildable designs.

The absence of established communication and coordination channels between designers and constructors at the design stage also limits effective implementation. Without structured platforms for collaboration, opportunities for early construction input are missed. Additionally, proposed changes by contractors are often ignored, either due to resistance from designers or concerns over design authority.

This discourages contractors from actively participating in design discussions and weakens collaborative efforts.

Institutional and systemic barriers also play a role. The lack of government support, in the form of policies, guidelines, or incentives, reduces industry-wide adoption of buildability assessment. Furthermore, the traditional procurement approach, which inherently separates design from construction, reinforces fragmented practices and discourages collaboration. Under such procurement systems, designers and contractors operate under separate contracts with limited incentives to cooperate.

Knowledge-related barriers are equally significant. Many designers have poor knowledge of construction operations, limiting their ability to foresee practical construction challenges. As a result, designers may rely on contractors to resolve buildability problems during construction, leading to inefficiencies, delays, and disputes. Compounding this issue is the fact that incentives are rarely provided to designers for developing more buildable designs, reducing motivation to prioritise buildability.

Other barriers include incomplete designs at the start of construction, which necessitate frequent changes and rework, and the difficulty in measuring or demonstrating the benefits of buildability assessment to the industry. When benefits are not clearly quantified, stakeholders may undervalue buildability assessment and resist its adoption.

Table 5.2 Barriers to buildability assessment implementation

S/No	*Barriers*
1	Poor co-ordination amongst different design disciplines
2	Lack of buildability assessment awareness
3	Design review pattern where construction personnel (builders) are invited to assess the building production information
4	Unwillingness to invest extra money and effort during the early stage of project
5	Designers value aesthetics and functionalities more than buildability in their designs
6	Design requirements on aesthetics often contradict buildability considerations
7	No established channel for co-ordination and communication between designers and constructors at the design stage
8	Lack of support from the government
9	Designers do not have adequate time to give sufficient attention to buildability
10	Designers rely on contractors to solve buildability problems
11	Traditional procurement approach inherently separates design from construction
12	Design requirements on functionalities often contradict buildability considerations
13	Designers' poor knowledge of construction operations
14	Incentives are not provided to designers for developing more buildable designs
15	Education and training on buildability are not provided to designers
16	Designers do not consider buildability important
17	Designers do not know the benefits of improved buildability of their designs/ difficulty in measuring the benefits of buildability to the industry
18	Proposed changes by contractors are usually ignored
19	Designs are incomplete when construction is started
20	Designers are not required to develop buildable designs

Source: Osuizugbo (2023).

In summary, the barriers to buildability assessment implementation arise from a combination of cultural, organisational, financial, technical, and institutional factors. As outlined in Table 5.2, these challenges are largely within the control of construction stakeholders. Addressing them requires increased awareness, education, improved coordination, supportive procurement systems, stronger government involvement, and a cultural shift toward collaboration. By overcoming these barriers, the construction industry can more effectively implement buildability assessment and realise its full potential in improving project delivery and performance.

5.8 Best Practice Case Studies

Best practice case studies play a critical role in demonstrating the practical value and effectiveness of buildability assessment in real construction projects. While theoretical discussions explain the concept and its potential benefits, case studies provide tangible evidence of how buildability assessment can be successfully implemented and the outcomes that can be achieved. They offer lessons learned, highlight success factors, and provide benchmarks that can guide stakeholders in adopting buildability assessment as a standard industry practice.

In the context of the construction industry, best practice case studies typically involve projects where buildability assessment was deliberately integrated into the design and planning stages, with active collaboration between clients, designers, and constructors. Such projects often show measurable improvements in cost control, construction duration, quality, and stakeholder satisfaction when compared with traditional project delivery approaches.

A common characteristic of best practice projects is the early involvement of constructors in the design process. In these cases, contractors or professional builders were engaged during the conceptual or schematic design stages to review production information and provide input on construction methods, sequencing, material selection, and site logistics. This early involvement enabled potential buildability issues to be identified and resolved before construction commenced, thereby reducing design changes, rework, and delays during execution.

5.8.1 Case Study: Early Contractor Involvement in UK Public Sector Construction

This is a real-world referenced case study illustrating buildability assessment (via early contractor involvement (ECI) and constructability practices) in the construction industry. This example is drawn from documented UK practice and widely accepted principles of buildability/constructability implementation.

Project Context: ECI is a collaborative procurement and project delivery approach widely used in the UK construction industry, particularly on complex public projects. ECI is designed to integrate contractor expertise early in the design and planning phases, enabling buildability (or constructability) issues to be identified and resolved before construction begins, thereby improving project outcomes. The UK Government's Highways Agency and Department for Transport pioneered ECI schemes in the early 2000s, embedding contractor participation in pre-construction design development as part of formal procurement strategies. These initiatives have been analysed through multiple case studies showing measurable benefits in cost, schedule, and risk management.

ECI Mechanism: In ECI-driven projects, a contractor is engaged before detailed design completion, typically after a selective tender process, to contribute construction expertise to design refinement, planning, sequencing, and risk mitigation. Under UK guidance on ECI, the contractor's role during the design phase includes:

- Advising on construction methods and logistics
- Reviewing design information for buildability issues
- Collaborating with designers to optimise plans
- Supporting realistic cost and programme forecasting
- Helping integrate buildable solutions into production documents

This process effectively operationalises buildability assessment by placing construction knowledge at the heart of design decisions, rather than deferring such input until after the design is complete.

Documented Outcomes from ECI Case Studies: Researchers have studied the first five UK government ECI projects to assess their success relative to traditional design-bid-build approaches. Key documented outcomes include:

Cost Savings: Across the first five ECI schemes implemented by the Highways Agency, total cost savings amounted to approximately £59.3 million, with an average saving of £5.4 million per scheme. This result highlights the value of early buildability assessment through contractor contributions that anticipate and reduce waste, rework, and unforeseen cost escalations.

Time Savings and Risk Management: Although time savings were less pronounced than cost benefits in these early cases, ECI consistently supported improved risk identification and planning, which mitigated cost impacts associated with late changes. For larger schemes, contractor involvement contributed to better programme planning and greater schedule reliability.

Enhanced Constructability: The involvement of contractors early in the design stage improved the quality of construction planning, helped refine build sequencing, and reduced the number of design-related queries and errors during execution. When constructors participated in design reviews, they brought practical insight into material handling, labour productivity, site logistics, and construction sequencing, effectively embodying buildability assessment principles.

Integrated Decision-Making: ECI's integrated team structure encourages open communication and shared ownership of design choices, aligning aesthetic, functional, and constructability objectives. This collaborative environment reduces clashes between design disciplines and supports more realistic, buildable solutions.

5.8.2 Lessons Learned and Best Practices

The documented UK case studies on ECI highlight several important lessons and best practices for the effective implementation of buildability assessments. A consistent finding across these projects is that early construction input is critical. Involving contractors at the earliest stages of a project, often during concept or preliminary design, enables practical construction expertise to inform design decisions before they become fixed. This early engagement helps identify buildability issues such as complex detailing, unrealistic sequencing, or impractical material choices, thereby reducing the likelihood of costly redesigns, delays, and rework during the construction phase.

Another key lesson is the importance of a collaborative team culture. Projects that successfully implemented ECI placed strong emphasis on cooperation between clients, designers, and contractors. Rather than operating in silos, these teams worked through regular design workshops, buildability reviews, and joint problem-solving sessions. Such collaborative practices promoted transparency, built trust, and ensured that design intent, construction constraints, and client objectives were aligned. As a result, decisions were made collectively, leading to more efficient, practical, and value-driven outcomes.

Systematic cost and risk analysis also emerged as a best practice. Integrating buildability assessments into established cost and risk management frameworks, such as target cost development, risk registers, and programme planning, allowed project teams to quantify the benefits of ECI more effectively. This approach helped justify early investment in contractor involvement by demonstrating savings through reduced risk, improved programme certainty, and enhanced value for money. Furthermore, continuous monitoring of cost and risk throughout the project lifecycle ensured that buildability improvements translated into measurable performance outcomes.

Finally, the procurement strategy was identified as a crucial enabler of successful ECI. Integrated procurement models, including ECI arrangements and two-stage tendering, provided contractors with both the authority and responsibility to influence design in a meaningful way. These models encouraged shared risk and reward, fostering innovation and constructive input from the supply chain. In contrast, traditional procurement routes that separate design and construction limited the contractor's ability to contribute early, making it more difficult to realise the full benefits of buildability assessment.

Overall, the case studies demonstrate that ECI is most effective when supported by early engagement, collaboration, structured cost and risk management, and appropriate procurement strategies.

The UK's experience with ECI illustrates a real-world, adopted model of buildability assessment in practice. By involving contractors early in design and decision-making, project teams can uncover and solve buildability challenges before they manifest on-site, leading to measurable improvements in cost efficiency, constructability, team collaboration, and risk mitigation. This documented practice reinforces the theoretical arguments for buildability assessment by showing that when buildability principles are actively applied through structured early involvement, tangible project benefits follow.

5.9 Chapter Conclusion

This chapter has established buildability assessment as a structured and practical mechanism for translating constructability principles into measurable project outcomes. While buildability is often discussed conceptually, its effective implementation depends on systematic evaluation processes that examine design decisions, construction methods, sequencing, and resource considerations before work begins on-site. By embedding buildability assessment into project workflows, stakeholders can identify potential inefficiencies, risks, and coordination issues early, thereby enhancing overall project performance.

The discussion highlighted the growing need for buildability assessment in an increasingly complex construction environment. Projects today are characterised by tight schedules, cost pressures, advanced technologies, and heightened regulatory demands. Without formal assessment mechanisms, opportunities to optimise construction processes may be overlooked, leading to rework, delays, and increased risk exposure. Buildability assessment addresses this gap by providing a proactive framework that evaluates the practicality and efficiency of design solutions in relation to real-world construction constraints.

This chapter further demonstrated that the benefits of buildability assessment extend beyond immediate project improvements. Direct benefits include reduced construction time, lower costs, improved quality, enhanced safety, and minimised rework. Indirect benefits emerge in the form of stronger collaboration, improved communication among stakeholders, and better risk management practices. Additional long-term advantages include organisational learning, knowledge transfer, and the development of industry-wide best practices. Together, these outcomes reinforce the value of integrating buildability assessment as a standard component of project governance.

Effective implementation, however, depends on the availability and quality of information. Comprehensive production information, accurate drawings and specifications, construction methodologies, and a clear understanding of applicable regulations are essential inputs for meaningful assessment. The reliability of the process is directly linked to the depth and clarity of project data, underscoring the importance of coordinated documentation and digital integration tools such as BIM to support analysis and visualisation.

This chapter also examined the practical realities of implementation, identifying both drivers and barriers within the construction industry. Drivers include client demand for efficiency, regulatory encouragement, competitive market pressures, and the growing adoption of collaborative procurement models. Conversely, barriers such as fragmented project structures, resistance to change, limited ECI, insufficient expertise, and perceived additional costs can hinder adoption. Addressing these challenges requires leadership commitment, capacity building, and a cultural shift toward proactive planning and interdisciplinary collaboration.

The best practice case study of ECI in UK public sector construction illustrated how integrating construction expertise during the design phase strengthens buildability outcomes. Lessons learned from such initiatives emphasise the importance of early engagement, transparent communication, structured assessment frameworks, and continuous feedback loops. These practices demonstrate that buildability assessment is most effective when embedded within collaborative project delivery models rather than treated as an isolated review exercise.

In conclusion, buildability assessment represents a critical bridge between design intent and construction reality. By formalising the evaluation of constructability considerations, it enhances decision-making, reduces uncertainty, and improves project outcomes. When supported by accurate information, strong stakeholder collaboration, and enabling organisational structures, buildability assessment becomes a powerful tool for driving efficiency, innovation, and sustainable performance across the construction industry.

Bibliography

Abdelhameed, W., & Ismaeil, E.M.H. (2024). BIM use to support the technical tasks in construction projects: A case study. *Buildings*, 14(6), 1534. https://doi.org/10.3390/buildings14061534

af Hällström, A., & Bosch-Sijtsema, P. (2019). Early contractor involvement in the construction industry: A preliminary literature review. In: Gorse, C., & Neilson, C.J. (Eds.) *Proceedings of the 35th annual ARCOM conference*, September 2–4, Leeds, Association of Researchers in Construction Management, pp. 314–323.

Afolabi, A., & Oyeyipo, O. (2017). The perception of future decision makers on the building profession. *Malaysian Construction Research Journal*, 21(1), 55–73.

Aina, O.O., & Wahab, A.B. (2011). An assessment of buildability problems in the Nigerian construction industry. *Global Journal of Research Engineering*, 11(2), 42–52.

Al Hamadani, S., Al Alawi, M., & Al Nuaimi, A. (2022). Constructability practices in construction industry in Muscat: Case study. *Asian Journal of Civil Engineering*, 23(7), 1141–1153. https://doi.org/10.1007/s42107-022-00475-3

Bamisile, A. (2004). *Building production management*. Foresight Press Ltd.

Bamisile, A. (2013). Buildability analysis of building projects. A paper presented at the 13th induction & investiture ceremony of the Council of Registered Builder of Nigeria (CORBON), November 20, 2013, Shehu Yar'adua Centre, Abuja.

Cheetham, D.W., & Lewis, J. (2001). Productivity, buildability and constructability: Is work study the missing link? In: Akintoye, A. (Ed.) *17th annual ARCOM conference*, September 5–7, University of Salford. Association of Researchers in Construction Management, 1, pp. 271–280.

Eadie, R., & Matthew, G. (2014). Analysing the advantages of early contractor involvement. *International Journal of Procurement Management*, 7(6), 661–676. https://doi.org/10.1504/IJPM.2014.064971

Faria, P.S., Sotelino, E., Carmo, C.S.T.D., & Nascimento, D.L.M. (2023). Evaluating construction projects' alternatives using lean construction and sustainability principles in an information model framework. *Sustainability*, 15(23), 16517. https://doi.org/10.3390/su152316517

High Speed Two. (2014). Early contractor involvement. Available at: https://assets.publishing.service.gov.uk/media/5a7dd855ed915d2ac884de4d/Early_contractor_involvement__ECI__guidance__Oct_2014_.pdf?utm_source=chatgpt.com (accessed December 21, 2025).

Hyde, R. (1995). Buildability as a design concept for architects: A case study of laboratory buildings. *Engineering, Construction and Architectural Management*, 2(1), 45–56. https://doi.org/10.1108/eb021002

Ihenketu, C., Iwudibiah, C.C., & Ebube, O.C. (2018). Assessment of impact of buildability and maintainability analysis on building delivery in Nigeria. *International Journal of Science and Advanced Innovative Research*, 3(3), 84–97.

Lappalainen, E., Uusitalo, P., Pikas, E., Seppänen, O., Peltokorpi, A., Uusitalo, P., Reinbold, A., & Menzhinskii, N. (2022). Improving design quality by contractor involvement: An Empirical study on effects. *Buildings*, 12(8), 1188. https://doi.org/10.3390/buildings12081188

Lam, P.T.I., & Wong, F.W.H. (2008). Implementing a buildability assessment model for buildability improvement. *Architectural Science Review*, 51(2), 173–184. https://doi.org/10.3763/asre.2008.5121

Lefebvre, F, & McAuley, B. (2019). An investigation into current procurement strategies that promote collaboration through early contractor involvement with regards to their suitability for Irish public work projects. In: *Proceedings of the 4th CitA BIM Gathering, Galway*, September 26, pp 209–221. https://doi.org/10.21427/rck5-e450

Low, S.P., & Abeyegoonasekera, B. (2000). Integrating buildability principles into ISO 9000 quality management systems. *Architectural Science Review*, 43(1), 45–56. 10.1080/00038628.2000.9696888

Low, S.P., & Abeyegoonasekera, B. (2001). Integrating buildability in ISO 9000 quality management systems: Case study of a condominium project. *Building and Environment*, 36(3), 299–312. https://doi.org/10.1016/S0360-1323(00)00004-4

Low, S.P., Liu, J.Y., & Oh, K.H. (2008). Influence of total building performance, spatial and acoustic concepts on buildability scores of facilities. *Facilities*, 26(1-Feb.), 85–104. https://doi.org/10.1108/02632770810840327

Ly, H. (2014). Buildability assessment model for building construction in Cambodia. *Chulalongkorn University Theses and Dissertations (Chula ETD)*. 69737. https://digital.car.chula.ac.th/chulaetd/69737

Moses, T., Heesom, D., Oloke, D., & Crouch, M. (2016). Exploring the benefits of early contractor involvement (Eci) and the mechanism for automated 5D BIM quantification process in offsite modular construction, 43–50. https://doi.org/10.29173/mocs6

Mydin, S.H., Zin, R.M., Majid, M.Z.A., Zahidi, M., & Memon, A.H. (2011a). Buildability attributes at design phase in Malaysian building construction. *International Journal of Sustainable Construction Engineering & Technology*, 2(1), 24–43.

Nourbakhsh, M., Mydin, S.H., Rosli, M.Z., Zolfagharian, S., Irizarry, J., & Zahidi, M. (2012). A conceptual model to assess the buildability of building structure at design stage in Malaysia. *Advanced Materials Research*, 446–449, pp. 3879–3884. https://doi.org/10.4028/www.scientific.net/amr.446-449.3879

Osuizugbo, I.C. (2020). The need for and benefit of buildability analysis: Nigeria as a case study. *Journal of Engineering, Design and Technology*, 19(5), 1207–1230. https://doi.org/10.1108/JEDT-08-2020-0338

Osuizugbo, I.C. (2023). Factors impeding the implementation of buildability assessment in building construction industry. *Journal of Engineering, Design and Technology*. https://doi.org/10.1108/JEDT-06-2022-0304

Osuizugbo, I.C. (2025). Buildability-led procurement: A new model for future construction projects delivery. *Construction Innovation: Information, Process, Management*. https://doi.org/10.1108/CI-10-2023-0272

Osuizugbo, I.C., & Ojelabi, R.A. (2020). Building production management practice in the construction industry in Nigeria. *Engineering Management in Production and Services*, 12(2), 56–73. https://doi.org/10.2478/emj-2020-0011

Osuizugbo, I.C., & Oshodi, O.S. (2023). Buildability attributes for improving the practice of construction management in Nigeria. *Built Environment Project and Asset Management*, 13(2), 270–289. https://doi.org/10.1108/BEPAM-03-2022-0051

Osuizugbo, I.C., Okolie, K.C., & Oshodi, O.S. (2022). Factors supporting the implementation of buildability assessment as a tool for buildability improvement. *Journal of Engineering, Design and Technology*. https://doi.org/10.1108/JEDT-12-2021-0738

Osuizugbo, I.C., Okolie, K.C., Oshodi, O.S., & Oyeyipo, O.O. (2023). Buildability in the construction industry: A systematic review. *Construction Innovation: Information, Process, Management*, 23(5), 1300–1322. https://doi.org/10.1108/CI-05-2022-0112

Pim-Wusu, M., Simpe, E.K., Adu Gyemfi, T., & Thwala, W.D. (2025). Understanding structural failures through buildability assessment: A case study of collapsed buildings in Accra, Ghana. *African Journal of Applied Research*, 11(5), 415–432. https://doi.org/10.26437/gby6m567

Procure Partnerships. (2025). How frameworks and early contractor involvement can reduce project costs. Available at: https://procurepartnerships.co.uk/news/how-frameworks-and-early-contractor-involvement-can-reduce-project-costs/?utm_source=chatgpt.com (accessed December 21, 2025).

Sanjaya, P.A., Joni, G.P., & Frederika, A. (2019). The role of contractor in improving buildability in construction projects in Bali. *MATEC Web of Conferences*, 276, p. 02023. https://doi.org/10.1051/matecconf/201927602023

Scape. (2025). Collaboration is one of the core values placed at the heart of SCAPE's suite of frameworks. Available at: https://scape.co.uk/news/how-can-early-contractor-involvement-eci-benefit-my-project?utm_source=chatgpt.com (accessed December 21, 2025).

Windapo, A.O., & Ogunsanmi, O.E. (2014). Evaluation of the barriers to the use of appropriate constructability practices on construction projects. *Journal of Construction Project Management and Innovation*, *4*(1), 734–754.

Wondimu, P., Klakegg, O., & Lædre, O. (2020). Early contractor involvement (ECI): Ways to do it in public projects. *Journal of Public Procurement* (ahead-of-print). https://doi.org/10.1108/JOPP-03-2019-0015

Wong, F.W.H., Lam, P.T.I., Chan, A.P.C., & Chan, E.H.W. (2006). A review of buildability performance in Hong Kong and strategies for improvement. *Surveying and Built Environment*, 17(2), 37–48.

Wong, F.W., Lam, P.T., Chan, E.H., & Shen, L (2007). A study of measures to improve constructability. *International Journal of Quality & Reliability Management*, 246, 586–601. https://doi.org/10.1108/02656710710757781

Zakaria, Z., Ismail, S., & Yusof, A.M. (2013). Buildability and maintainability in final account closing: The greatest challenge for construction industry. *Conference Paper*, March 2013, University of Technology Malaysia, Kuala Lumpur. https://doi.org/10.13140/RG.2.1.1758.2806

6 Buildability in Health and Safety

6.1 Health and Safety in the Construction Industry: An Overview

The construction industry plays a vital role in economic development and infrastructure provision worldwide; however, it is consistently identified as one of the most hazardous industrial sectors in terms of occupational health and safety (OHS). Construction activities involve complex interactions between people, materials, machinery, and the physical environment, often under time pressure and changing site conditions. These characteristics contribute to a heightened risk profile when compared to manufacturing or service-based industries. Despite significant regulatory developments and advances in safety management practices, construction continues to record disproportionately high rates of accidents, injuries, and fatalities globally.

6.1.1 Nature and Characteristics of Construction Work

Construction work is inherently dynamic, temporary, and complex, distinguishing it from many other industrial sectors. Each construction project is unique, characterised by distinct site conditions, design requirements, environmental constraints, and organisational arrangements. Unlike manufacturing environments, where production processes are standardised, repetitive, and conducted in controlled settings, construction sites evolve continuously as work progresses through different phases such as excavation, structural framing, services installation, and finishing. This constantly changing nature creates a moving risk landscape, where hazards emerge, disappear, and re-emerge as activities change, making systematic hazard identification and risk control particularly challenging.

The temporary nature of construction sites further compounds safety challenges. Workplaces are established for limited durations and then dismantled once the project is completed. As a result, permanent safety infrastructure is rarely feasible, and reliance is placed on temporary works such as scaffolding, formwork, edge protection, and access systems. These temporary arrangements must frequently be altered or relocated, increasing the potential for errors, inadequate installation, or misuse. In addition, construction work is often undertaken outdoors, exposing workers to

DOI: 10.1201/9781003747284-8

weather conditions such as heat, rain, wind, and poor visibility, all of which can adversely affect safety performance and decision-making.

Another defining characteristic of construction work is the diversity of tasks performed within a confined space and timeframe. Multiple trades often operate simultaneously, each with different tools, materials, and work methods. This overlap can result in congestion, conflicting activities, and increased exposure to hazards such as falling objects, vehicle movements, and noise. Coordinating these activities requires careful planning and communication; when such coordination is lacking, the likelihood of accidents increases significantly.

Construction projects also involve a highly fragmented organisational structure. Numerous stakeholders, including clients, designers, main contractors, subcontractors, and suppliers, contribute to project delivery, often under separate contractual arrangements. While this fragmentation allows for specialisation and flexibility, it can also lead to blurred lines of responsibility and reduced accountability for health and safety. Communication failures between parties are a common contributing factor in construction accidents, particularly when safety information is not effectively shared across organisational boundaries.

Subcontracting is a widespread practice in construction and plays a significant role in shaping site safety. Subcontractors often employ workers on short-term or project-based contracts, resulting in a workforce with varying levels of experience, training, and familiarity with site-specific hazards. Language barriers, cultural differences, and differing safety standards between organisations can further complicate safety management. These conditions place additional demands on-site supervision and safety coordination to ensure consistent standards are maintained.

Collectively, the dynamic, temporary, and fragmented nature of construction work creates an environment in which hazards are difficult to predict and control using traditional safety approaches alone. These characteristics underscore the need for proactive, design-led, and systems-based strategies, such as improved buildability, to address health and safety risks at their source rather than relying solely on-site-level controls.

6.1.2 Accident Statistics and Global Health and Safety Performance

The construction industry continues to record some of the highest rates of occupational accidents and fatalities worldwide, making health and safety a persistent global concern. According to the International Labour Organization, construction workers comprise only about 7–10% of the global workforce yet account for approximately 30–40% of all workplace fatalities. This disproportionate representation highlights the inherently hazardous nature of construction activities and the ongoing challenges faced in managing risks effectively. Notably, this trend is evident across both developed and developing economies, indicating that economic advancement alone does not eliminate construction safety risks.

In many countries, construction consistently ranks among the top sectors for fatal occupational accidents. Developing nations often experience higher accident rates due to weaker regulatory frameworks, limited enforcement, and inadequate training. However, even in highly regulated environments, such as Europe, North America, and Australia, construction fatalities remain a major concern. This suggests that regulatory compliance alone is insufficient and must be complemented by proactive risk management and design-led safety strategies.

Analysis of accident data reveals a consistent pattern in the types of incidents occurring on construction sites. Falls from height are widely recognised as the leading cause of death in the industry, frequently associated with roof work, scaffolding, ladders, and unprotected edges. These incidents are often linked to inadequate access planning, insufficient edge protection, and poorly designed or installed temporary works. Other common causes of fatal and serious injuries include being struck by moving vehicles or falling objects, electrocution from live services, trench collapses, and accidents involving heavy machinery and equipment.

In addition to fatal accidents, non-fatal injuries, and occupational health issues impose a substantial burden on the construction workforce. Musculoskeletal disorders resulting from manual handling, repetitive tasks, and awkward postures are widespread, while cuts, fractures, and crush injuries remain common. Long-term occupational illnesses, such as respiratory diseases caused by dust exposure and hearing loss due to prolonged noise exposure, further affect workers' quality of life and employability. These health issues often receive less attention than acute injuries but contribute significantly to lost working days and long-term disability.

The consequences of poor health and safety performance extend beyond individual workers to organisations and society as a whole. Construction accidents result in project delays, disrupted workflows, and reduced productivity. Financial impacts include increased insurance premiums, compensation claims, legal costs, and potential regulatory penalties. Reputational damage following serious incidents can also affect a company's ability to secure future work. Consequently, improving health and safety performance in construction is not only a moral and legal responsibility but also a strategic business imperative that directly influences project success and organisational sustainability.

6.1.3 Regulatory Frameworks and Legal Responsibilities

In response to persistently high accident and fatality rates in the construction industry, governments across the world have developed comprehensive OHS regulatory frameworks aimed at protecting workers and reducing workplace risks. These regulations generally impose legal duties on employers to ensure, so far as is reasonably practicable, the health, safety, and welfare of workers. Core requirements typically include the provision of safe systems of work, adequate training and supervision, systematic risk assessments, and the supply and correct use of personal protective equipment (PPE). Compliance with these obligations forms the foundation of construction health and safety management in most jurisdictions.

In recent decades, regulatory frameworks have evolved to recognise that responsibility for construction safety extends beyond contractors and site management. Increasingly, clients and designers are assigned explicit legal duties to consider health and safety throughout the entire project lifecycle, from concept design to construction and eventual maintenance. This shift reflects growing recognition that many construction hazards originate from early design decisions and project planning choices, rather than solely from site activities.

A prominent example of this approach is the United Kingdom's Construction (Design and Management) Regulations (CDM), which place specific duties on clients, designers, and contractors to manage health and safety risks collaboratively. Under the CDM Regulations, designers are legally required to eliminate foreseeable risks where possible and, where elimination is not practicable, to reduce risks through informed design decisions. This includes considering safe access, construction sequencing, maintenance requirements, and the use of materials and systems that minimise risk. Similar regulatory principles are evident in construction safety legislation in countries such as Australia, Singapore, and several European Union member states, where there is increasing emphasis on proactive, upstream risk management rather than reliance on-site–based controls alone.

Despite the existence of robust legislative frameworks, compliance with construction health and safety regulations remains inconsistent across the industry. Larger organisations often have dedicated safety professionals, formal management systems, and greater capacity to implement regulatory requirements. In contrast, small and medium-sized contractors frequently face significant challenges, including limited financial resources, insufficient safety expertise, and high workforce turnover. These constraints can result in superficial compliance, where safety documentation exists but is not effectively implemented on-site.

Enforcement also varies between jurisdictions, with limited inspection capacity and inconsistent penalties undermining the effectiveness of regulations in some regions. Consequently, legislation alone is insufficient to drive sustained improvements in construction health and safety performance. This has led to growing recognition of the need for complementary approaches that integrate safety into design, planning, and decision-making processes. Embedding safety through buildability-focused design and collaborative project delivery offers a means of reinforcing regulatory objectives and achieving more consistent and meaningful health and safety outcomes.

6.1.4 Traditional Approaches to Construction Health and Safety

Traditionally, health and safety management in the construction industry has been centred on-site–level controls aimed at preventing accidents during the construction phase. These controls typically include worker induction programmes, safety training, toolbox talks, the use of PPE, method statements, risk assessments, and on-site supervision. Such measures are designed to manage hazards once construction activities are underway and to ensure that workers follow established safety procedures. While these practices are essential components of any safety management system, they largely represent a reactive approach to risk control, addressing hazards after they have already been introduced into the construction process.

A defining characteristic of traditional safety approaches is their strong reliance on worker behaviour and compliance. Safety performance is often measured by adherence to rules, the correct use of PPE, and the avoidance of unsafe acts. When accidents occur, investigations frequently focus on individual behaviour, attributing incidents to carelessness, lack of attention, or failure to follow procedures. This behavioural emphasis has led to the widespread perception that accidents are primarily caused by human error.

However, contemporary safety research challenges this view, arguing that human error is rarely the sole or fundamental cause of accidents. Reason's (1997) work on organisational accidents demonstrates that unsafe acts are often the final outcome of deeper systemic failures. Factors such as inadequate design, unrealistic project schedules, insufficient planning, and poor coordination between stakeholders create conditions in which errors are more likely to occur. In such environments, even well-trained and experienced workers may struggle to perform tasks safely.

Traditional approaches also tend to over-rely on administrative controls and PPE, which occupy the lower levels of the hierarchy of risk control. These measures are inherently less effective than hazard elimination or substitution because they depend on consistent human behaviour and supervision. For example, requiring workers to use fall-arrest harnesses addresses the risk of falling but does not eliminate the hazard itself. If access systems are poorly designed or work is rushed, the likelihood of non-compliance increases, reducing the effectiveness of such controls.

As a result of these limitations, there has been growing criticism of safety strategies that focus exclusively on-site–based practices while neglecting upstream influences. Researchers and practitioners increasingly recognise that many hazards are embedded in project decisions made during the design and planning stages. Design complexity, constrained site layouts, and unrealistic construction programmes can significantly undermine site safety, regardless of the quality of on-site controls.

This evolving understanding has prompted a shift toward more proactive and preventative approaches to construction health and safety. By examining how early project decisions influence risk, the industry is beginning to move beyond traditional reactive models toward integrated strategies that address safety at its source, laying the foundation for concepts such as buildability and design-led safety improvement.

6.1.5 Design-Related Causes of Construction Accidents

A growing body of research has demonstrated that a significant proportion of construction accidents can be traced back to decisions made during the design and planning stages of a project, long before construction activities begin on-site. Traditional safety management approaches have often treated design and construction as separate domains, with safety responsibility largely assigned to contractors during the construction phase. However, empirical evidence increasingly challenges this separation, highlighting the critical influence of design-related factors on construction health and safety outcomes.

Haslam et al. (2005), in a landmark investigation of construction accidents in the United Kingdom, identified design as a major contributing factor in many incidents. Their study revealed that inadequate consideration of safe access, construction sequencing, maintenance requirements, and temporary works frequently created hazardous conditions on-site. For example, designs that failed to provide adequate space for safe access systems or lifting operations forced contractors to improvise, often leading to unsafe work practices. Similarly, neglecting maintenance considerations during design exposed workers to risks during both construction and the operational life of the facility.

Overly complex and poorly coordinated designs further exacerbate safety risks. Complex geometries, bespoke details, and intricate interfaces between structural, architectural, and services elements often require non-standard construction methods. These methods can involve increased manual handling, extended working at height, or work in confined spaces, all of which elevate accident risk. Poor design coordination may also result in clashes between building elements, leading to rework and unplanned modifications on-site. Such conditions frequently create time pressure, confusion, and frustration among workers, increasing the likelihood of errors and accidents.

Insufficient design detail is another common contributor to construction accidents. When drawings and specifications lack clarity or fail to address construction sequencing and tolerances, contractors are left to make assumptions or decisions under pressure. This can result in unsafe assembly sequences, unstable temporary conditions, or inadequate support during critical stages of construction. Rushed corrective actions and rework further expose workers to hazards, particularly when changes are implemented without proper risk assessment.

These findings fundamentally challenge the traditional view that construction safety can be managed effectively through site-level controls alone. While site safety measures remain essential, they are often insufficient to counteract hazards embedded in poor design. Instead, designers play a crucial role in shaping the risk profile of a project by influencing construction methods, site layout, and sequencing. Recognising this responsibility underscores the need for greater integration between design and construction teams and supports the adoption of proactive, design-led approaches to health and safety management.

6.1.6 Health and Safety Culture in Construction

Health and safety culture refers to the collective values, beliefs, attitudes, and behaviours that determine how safety is perceived, prioritised, and managed within an organisation or industry. In the construction sector, safety culture plays a critical role in shaping day-to-day practices and influencing health and safety performance. Unlike formal safety systems, which focus on rules, procedures, and documentation, safety culture reflects how safety is enacted in practice. It determines whether policies are actively followed or merely complied with at a superficial level.

Safety culture in construction varies significantly across organisations, projects, and geographical regions. A positive safety culture is typically characterised by visible leadership commitment, effective communication, worker involvement in safety decision-making, and a willingness to learn from incidents and near-misses. When managers and supervisors consistently demonstrate that safety is a priority, through their actions as well as their words, workers are more likely to adopt safe behaviours and engage proactively in risk management. Open communication channels also enable workers to report hazards and near-misses without fear of blame, allowing risks to be addressed before serious accidents occur.

Despite its importance, the construction industry has frequently been criticised for a production-oriented culture that places greater emphasis on cost, time, and productivity than on safety. Competitive tendering, tight project schedules, and financial pressures can encourage shortcuts, risk-taking, and the normalisation of unsafe practices. In such environments, workers may feel pressured to prioritise output over safety, while supervisors may overlook unsafe behaviour to maintain progress. This culture can discourage the reporting of incidents and near-misses, limiting opportunities for learning and improvement.

A production-driven culture can also undermine formal safety management systems. Even where comprehensive safety policies and procedures exist, their effectiveness depends on consistent implementation and enforcement. If safety is perceived as a bureaucratic requirement rather than a core organisational value, compliance may become superficial. This disconnect between formal systems and actual practice is a common feature of poor safety performance in construction.

Improving health and safety culture requires sustained leadership commitment at all levels of the project organisation, including clients, designers, and contractors. Clients, in particular, influence safety culture by setting project priorities and expectations. When safety is treated as a core project value rather than merely a regulatory obligation, it is more likely to be integrated into decision-making processes. Designers also play a key role by considering safety and buildability during the design phase, thereby reinforcing the message that safety is a shared responsibility. Ultimately, a strong safety culture supports proactive risk management and creates the conditions necessary for meaningful and lasting improvements in construction health and safety performance.

6.1.7 Occupational Health Issues in Construction

While construction health and safety discourse often focuses on accidents and fatalities, occupational health issues represent an equally significant but frequently overlooked challenge within the industry. Construction workers are routinely exposed to a wide range of health hazards arising from the nature of construction activities and work environments. These hazards include airborne dust such as silica and asbestos, excessive noise from machinery and tools, vibration from handheld equipment, hazardous chemicals, and prolonged exposure to extreme weather conditions. Repeated or long-term exposure to these hazards can result in chronic and sometimes irreversible health conditions.

Respiratory diseases are among the most serious occupational health issues in construction, commonly linked to inhalation of fine dust particles generated during cutting, grinding, and demolition activities. Conditions such as silicosis, chronic obstructive pulmonary disease, and occupational asthma may develop gradually and often become apparent only after prolonged exposure. Similarly, prolonged exposure to high noise levels can lead to noise-induced hearing loss, which is typically permanent and can significantly affect quality of life. Hand–arm vibration syndrome, caused by extended use of vibrating tools, is another prevalent occupational illness that can result in pain, numbness, and loss of function.

Musculoskeletal disorders are also widespread in construction and are closely associated with manual handling, repetitive tasks, awkward postures, and physically demanding work. These conditions contribute substantially to lost working days, reduced productivity, and long-term disability. Unlike acute injuries, occupational health conditions often develop slowly and may not be immediately linked to workplace exposure, making them more difficult to detect, monitor, and manage effectively.

One of the key challenges in addressing occupational health risks is their relative invisibility compared to accidents. Construction safety management has traditionally prioritised immediate and visible risks, such as falls and struck-by incidents, while long-term health effects receive less attention. Workers may also underreport health symptoms due to job insecurity, lack of awareness, or the delayed onset of illness. Consequently, occupational health risks are often under-prioritised in both project planning and day-to-day site management.

Design decisions play a critical role in influencing occupational health outcomes. Buildability-oriented design can reduce exposure to health hazards by minimising manual handling through prefabrication, selecting materials with lower dust or chemical emissions, and designing construction methods that reduce noise and vibration. Additionally, considering weather protection, ergonomics, and safe maintenance access during design supports long-term worker well-being.

These considerations reinforce the need for a holistic approach to construction health and safety, one that integrates accident prevention with proactive occupational health management. Addressing both immediate and long-term risks is essential for protecting workers, improving productivity, and achieving sustainable health and safety performance across the construction industry.

6.1.8 Emerging Trends and the Shift Toward Proactive Safety

In recent decades, the construction industry has experienced a significant shift in health and safety thinking, moving away from predominantly reactive approaches toward proactive and preventative strategies. This change has been driven by persistent accident rates, advances in safety research, and growing recognition that traditional site-based controls alone are insufficient to achieve substantial and sustained improvements. Emerging safety concepts increasingly focus on addressing risks at their source, rather than managing hazards only after they appear on construction sites.

Central to this shift are approaches such as Prevention through Design (PtD), Design for Safety (DfS), and Safe by Design. These concepts promote the systematic identification and elimination of hazards during the early design stages of a project, when the greatest opportunity exists to influence construction methods and risk exposure. By integrating safety considerations into design decisions, these approaches seek to reduce reliance on administrative controls and PPE, which are inherently less effective due to their dependence on human behaviour and supervision.

The move toward proactive safety aligns closely with buildability principles. Buildability encourages designs that are efficient, logical, and easy to construct, thereby reducing complexity and uncertainty during construction. Simplified designs, standardised components, and well-planned construction sequences minimise the need for high-risk activities such as extensive work at height, heavy manual handling, and complex temporary works. In this way, buildability-focused design directly contributes to hazard elimination and risk reduction, supporting the objectives of PtD and DfS.

Technological advancements have further reinforced this proactive shift. Tools such as Building Information Modelling enable designers and contractors to visualise construction processes, identify potential hazards, and test alternative design solutions before work begins. These digital capabilities support early collaboration between project stakeholders and facilitate informed decision-making regarding safety and buildability. As a result, risks can be addressed proactively rather than discovered through incidents or near-misses on-site.

Importantly, this emerging approach reflects a broader change in how health and safety is perceived within construction projects. Safety is increasingly viewed not as a standalone compliance function but as an integral component of overall project performance, alongside cost, time, and quality. When safety considerations are embedded within design and planning processes, they often lead to improved efficiency, reduced rework, and better productivity outcomes.

This proactive, design-led perspective represents a fundamental cultural and organisational shift within the construction industry. By integrating health and safety with buildability and project delivery objectives, the industry is better positioned to achieve sustainable improvements in safety performance and protect workers throughout the project lifecycle.

In summary, the construction industry continues to face significant health and safety challenges due to the inherently hazardous nature of its work, fragmented organisational structures, and traditional reliance on reactive safety measures. High accident and fatality rates persist despite extensive regulatory frameworks, highlighting the limitations of site-focused approaches. Research increasingly demonstrates that many safety risks originate from design and planning decisions, underscoring the importance of addressing health and safety upstream. This evolving understanding provides a strong foundation for examining the role of buildability in improving health and safety outcomes, which is further explored in subsequent sections of this chapter.

6.2 Conceptual Link: Buildability and Health and Safety

The relationship between buildability and health and safety in construction is increasingly recognised as fundamental to improving project outcomes and reducing occupational risks. Buildability, sometimes used interchangeably with constructability, refers to the extent to which the design of a project facilitates efficient, safe, and economical construction. Health and safety, meanwhile, encompasses the prevention of accidents, injuries, occupational diseases, and fatalities throughout the project lifecycle. The conceptual link between these two domains lies in the premise that design decisions directly shape construction methods, site conditions, and worker exposure to hazards. Consequently, buildability acts as a critical upstream determinant of health and safety performance.

6.2.1 Evolution of Buildability and Safety Thinking

Historically, the concepts of buildability and health and safety developed along largely separate trajectories within construction projects. Design teams traditionally prioritised architectural aesthetics, functional requirements, and structural performance, with limited consideration of how designs would be executed on-site. Health and safety, by contrast, was viewed primarily as the responsibility of contractors during the construction phase, to be managed through site supervision, method statements, and compliance with regulations. This division of responsibility created a disconnect between design intent and construction reality.

Traditional procurement routes, particularly the design–bid–build model, reinforced this separation. Under this approach, designers typically completed most of the design before contractors were appointed, leaving little opportunity for construction expertise to inform design decisions. As a result, designs were often developed without adequate consideration of practical construction methods, site constraints, or worker safety. Contractors were then required to adapt these designs during construction, sometimes resorting to complex or unsafe work practices to meet project requirements.

Persistent high accident and fatality rates across the construction industry prompted increasing scrutiny of this traditional model. Researchers and safety professionals began to question the prevailing assumption that accidents were primarily caused by unsafe worker behaviour. Influential studies demonstrated that many incidents could be traced back to upstream decisions related to design, planning, and project organisation. Reason's (1997) work on organisational accidents, in particular, highlighted that unsafe acts are often symptoms of deeper systemic failures rather than isolated individual errors.

This evolving understanding led to a paradigm shift in safety thinking, recognising that health and safety performance is shaped by the entire project system rather than by site activities alone. Design complexity, inadequate planning, unrealistic schedules, and poor coordination were increasingly identified as root causes of unsafe conditions. In this context, the role of designers in influencing construction risk began to receive greater attention, challenging the long-standing division between design and construction safety responsibilities.

Buildability emerged as a unifying concept within this shift in thinking. By focusing on how design decisions affect construction processes, buildability integrates considerations of efficiency, cost, quality, and safety into a single framework. Rather than treating safety as an add-on or compliance issue, buildability encourages designers to consider construction methods, sequencing, access, and worker safety from the earliest stages of project development. This integrated approach represents a significant evolution in construction practice, aligning design intent with safe and efficient construction outcomes and laying the foundation for more proactive, design-led health and safety management.

6.2.2 Design as a Determinant of Construction Risk

The conceptual link between buildability and health and safety is founded on the recognition that construction risk is largely shaped during the early design stages of a project. Decisions made at this point determine construction methods, site layout, sequencing of activities, and the extent to which workers are exposed to hazards. As a result, many risks are effectively "designed in" or "designed out" long before construction begins. This perspective represents a significant departure from traditional views that regard safety risks as issues to be managed solely during the construction phase.

Empirical research strongly supports the influence of design on construction safety outcomes. Haslam et al. (2005) demonstrated that a substantial proportion of construction accidents could be traced to design-related factors, including inadequate consideration of safe access, construction sequencing, temporary works, and maintenance requirements. Designs that fail to address these aspects often create hazardous site conditions that cannot be fully mitigated through site-based controls alone. These findings highlight that construction risk is not merely a function of worker behaviour or on-site supervision but is deeply influenced by upstream decisions made by designers and planners.

Poor buildability is a key mechanism through which design-related risks arise. Designs characterised by complex geometries, bespoke components, and limited tolerances frequently require complicated construction methods. Such methods may involve extensive working at height, heavy manual handling, or operations in confined spaces, all of which significantly increase accident risk. In addition, congested workspaces resulting from inadequate spatial planning can lead to trade overlap, restricted movement, and increased potential for struck-by incidents. Under these conditions, safety management becomes heavily reliant on procedural controls, detailed method statements, and strict worker compliance, which are inherently less reliable than hazard elimination.

Conversely, designs that prioritise buildability have a demonstrable positive impact on health and safety performance. Buildable designs simplify construction processes by promoting logical sequencing, standardised components, and clear access arrangements. By reducing complexity and uncertainty, such designs minimise the need for high-risk activities and reduce exposure to hazards. For example, incorporating prefabricated elements can significantly decrease working at height and manual handling, while designing for adequate access enables safer installation and maintenance activities.

This contrast illustrates that buildability extends beyond considerations of efficiency and cost; it is a fundamental determinant of construction health and safety outcomes. By shaping the nature and extent of risks encountered on-site, design decisions play a critical role in either exacerbating or mitigating hazards. Recognising design as a primary determinant of construction risk underscores the importance of integrating buildability and safety considerations early in the project lifecycle, reinforcing the need for proactive, design-led approaches to health and safety management.

6.2.3 PtD and DfS

The conceptual connection between buildability and health and safety is strongly reinforced through the principles of PtD and DfS. These approaches represent a proactive shift in construction safety management, advocating for the elimination or mitigation of hazards during the design phase rather than relying solely on administrative controls, PPE, or reactive measures implemented during construction. By addressing safety at its source, PtD and DfS aim to reduce both the likelihood and severity of accidents, creating safer work environments while also supporting project efficiency and quality.

PtD aligns closely with buildability principles because it explicitly requires designers to consider how their decisions influence construction methods, site conditions, and worker exposure to hazards. For instance, selecting construction systems that minimise temporary works, enable ground-level assembly, or reduce manual lifting directly decreases the number of high-risk activities on-site. Similarly, designing for modularity and prefabrication not only improves construction efficiency but also limits the duration and complexity of tasks that involve elevated work or awkward handling. In this sense, buildability provides the practical mechanism through which PtD objectives can be operationalised, translating design intent into tangible reductions in risk.

DfS further reinforces this approach by embedding hazard identification and risk reduction into standard design processes. It encourages multidisciplinary collaboration, ensuring that input from contractors, safety professionals, and clients informs design decisions. This collaboration allows designers to anticipate site constraints, optimise sequencing, and plan for safe access and egress, all of which contribute to reducing accidents and occupational health risks. DfS also emphasises the importance of considering maintenance and long-term operational safety during design, extending the benefits of proactive hazard management beyond the construction phase.

Regulatory frameworks in several countries have formalised the link between design and safety, assigning explicit legal responsibilities to designers. For example, the United Kingdom's Construction (Design and Management) Regulations (CDM) require designers to eliminate or reduce foreseeable risks "so far as is reasonably practicable" through their design choices. These regulations recognise that designers occupy a pivotal position in shaping construction risk and that proactive safety considerations must be integrated into the earliest project stages. Similar principles are embedded in legislation in Australia, Singapore, and parts of the European Union, reflecting a global shift toward upstream risk management.

By embedding PtD and DfS principles into practice, buildability-focused design becomes a central tool for enhancing health and safety performance. It moves the industry away from reliance on procedural controls and PPE, which are inherently dependent on human compliance, and towards systemic, design-led solutions that prevent hazards from arising in the first place. This approach demonstrates that safety, efficiency, and quality are not competing objectives but can be mutually reinforced when considered holistically during the design process. Ultimately, PtD and DfS highlight that proactive, buildability-informed design is a critical determinant of construction safety and a key mechanism for achieving sustainable improvements in health and safety outcomes.

6.2.4 Buildability as a Systemic Risk Control

A central dimension of the conceptual link between buildability and health and safety lies in its role as a systemic risk control mechanism. Traditional construction safety measures, such as PPE, warning signs, training programmes, and method statements, typically occupy the lower levels of the hierarchy of risk control. While these interventions are important, their effectiveness is inherently limited because they depend on consistent human compliance and supervision, both of which are prone to error. In contrast, buildability-focused interventions operate at the top of the hierarchy by addressing hazards at their source, thereby reducing or eliminating the underlying risks before they manifest on-site. This proactive approach moves safety management from reactive control toward systemic prevention.

Prefabrication and off-site assembly provide clear examples of how buildability functions as a systemic risk control. By manufacturing components in controlled factory environments, many high-risk activities associated with traditional on-site construction, such as working at height, handling heavy materials in confined spaces, and exposure to adverse weather, are either significantly reduced or entirely eliminated. This not only lowers the probability of accidents but also creates safer, more predictable construction conditions. Similarly, simplifying structural connections, reducing the number of bespoke components, and standardising assembly methods mitigate the risk of errors during installation, while also facilitating safer sequencing and access for workers. These design-led strategies exemplify the potential of buildability to integrate safety directly into project planning and execution.

From a theoretical perspective, the systemic role of buildability aligns closely with Reason's (1997) "Swiss cheese model" of accident causation. This model posits that accidents rarely result from a single unsafe act; instead, they arise from a combination of latent conditions, active failures, and gaps in organisational defences. In this framework, poor buildability constitutes a latent condition, a hidden design-related weakness that creates vulnerabilities in the system. When combined with other latent factors, such as tight schedules, inadequate supervision, or inexperienced workers, these gaps can align and result in accidents. By addressing buildability issues early, designers can remove or reduce these latent conditions, effectively "closing the holes" in the safety system before construction begins.

Moreover, treating buildability as a systemic control emphasises the integration of safety with other project objectives, such as constructability, cost efficiency, and quality. When designs are logical, standardised, and construction-friendly, hazards are naturally reduced, project performance improves, and reliance on lower-level, behaviour-dependent safety measures diminishes. This approach reframes safety not as an isolated concern addressed solely through compliance, but as an inherent attribute of design quality and project planning.

In summary, buildability represents a top-tier, systemic risk control strategy that transforms construction safety management. By embedding hazard reduction into design and planning, buildability mitigates latent risks, reduces dependence on reactive site measures, and aligns with broader organisational and project objectives. This perspective underscores the critical importance of integrating buildability considerations into early project stages to achieve sustainable improvements in construction health and safety outcomes.

6.2.5 Integration of Buildability into Project Lifecycle

The conceptual relationship between buildability and health and safety achieves its full potential when considered across the entire project lifecycle, from initial design through construction and into operation and maintenance. Early-stage decisions, such as the selection of structural systems, materials, and construction methodologies, have profound and lasting implications for worker safety, project efficiency, and long-term occupational health. Integrating buildability into the lifecycle ensures that safety is not treated as an isolated or reactive concern but as an inherent aspect of design and project planning.

During the design phase, decisions regarding the choice of structural systems can significantly affect construction risk. For instance, opting for modular or prefabricated components can reduce on-site working at height, manual handling, and exposure to adverse environmental conditions. Similarly, designs that minimise complex geometries and allow straightforward sequencing enable safer construction practices, reducing reliance on procedural controls and the likelihood of error. By embedding these considerations into design, buildability becomes a proactive mechanism for risk elimination rather than a post-hoc mitigation strategy.

Buildability considerations also extend beyond immediate construction activities to influence long-term operational safety. For example, the incorporation of permanent access systems for façade maintenance, roof inspection, or plant equipment significantly reduces occupational risks throughout the building's lifecycle. While these features improve safety for maintenance personnel in the operational phase, they also facilitate safer installation during construction, illustrating the dual benefits of lifecycle-oriented design. Likewise, selecting durable, lightweight, and easy-to-handle materials not only improves installation efficiency but also reduces long-term handling risks and potential for structural degradation that could create safety hazards over time.

Construction sequencing is another critical area where lifecycle-integrated buildability enhances safety. Thoughtful design that anticipates trade interactions, material delivery logistics, and site layout reduces congestion, overlapping activities, and conflict between trades, all common contributors to accidents. By planning the sequence of work during the design stage, designers can minimise the need for temporary works, reduce hazardous work conditions, and enable more predictable, safer workflows. This lifecycle perspective shifts health and safety management upstream, emphasising prevention rather than reliance on reactive, site-level interventions.

Furthermore, integrating buildability into lifecycle planning encourages collaboration among project stakeholders, including clients, designers, contractors, and safety professionals. Early engagement facilitates the identification of risks that may not be immediately apparent to designers alone and allows practical construction constraints to inform design decisions. This collaborative approach not only improves safety but also enhances project efficiency, reduces rework, and aligns with broader goals of quality and sustainability.

In conclusion, the integration of buildability into the project lifecycle ensures that health and safety considerations are embedded from design through operation. By addressing hazards proactively, facilitating safe sequencing, providing permanent access solutions, and promoting collaboration, buildability transforms safety management into a continuous, system-wide process. This holistic approach demonstrates that health and safety performance is intrinsically linked to design quality and project planning, extending benefits far beyond the construction phase into the operational life of the asset.

6.2.6 Role of Stakeholders in Buildability-Driven Safety

The link between buildability and health and safety highlights the central role of stakeholders in ensuring that design and construction decisions actively reduce risks. Achieving buildability-driven safety is not the responsibility of any single party; it requires the collective involvement of designers, clients, contractors, subcontractors, and safety professionals. Each stakeholder brings unique expertise and perspectives, and effective collaboration is essential for identifying hazards early, implementing practical solutions, and ensuring that safety is integrated throughout the project lifecycle.

Designers hold a key position in shaping the constructability and inherent safety of a project. Their decisions regarding structural systems, materials, access provisions, and construction sequencing directly influence the risks faced by workers. However, designers may lack practical insights into the challenges of construction, particularly in complex or unconventional projects. This is where early contractor involvement becomes invaluable. Contractors bring hands-on experience, knowledge of site conditions, and understanding of construction methods, enabling them to provide feedback on potential hazards and suggest safer, more buildable solutions. Involving contractors during the design phase also reduces the likelihood of unsafe workarounds or rework during construction, which are common sources of accidents.

Clients, meanwhile, exert a significant influence through project objectives, procurement strategies, and decision-making processes. When clients prioritise cost and time above buildability and safety, designers and contractors may be pressured to adopt complex, high-risk solutions or forego safer alternatives. Such constraints can compromise both construction efficiency and worker safety. Conversely, clients who emphasise safety as a core project value, and who support integrated and collaborative delivery models, create an environment where buildability-led improvements can thrive. By promoting early engagement of all relevant stakeholders, allocating sufficient time for design development, and recognising the long-term benefits of safer construction practices, clients play a pivotal role in enabling proactive risk management.

Safety professionals also contribute to the buildability-driven safety process by providing expertise in hazard identification, risk assessment, and compliance with regulatory standards. Their involvement during design and planning ensures that safety is considered systematically and not as an afterthought. Moreover, safety professionals help translate design concepts into practical safety measures, bridging the gap between design intent and on-site implementation.

The interaction between these stakeholders fosters a culture of shared responsibility and proactive safety management. When designers, contractors, clients, and safety experts collaborate effectively, buildability principles can be integrated into design decisions, construction methods can be optimised for safety, and latent hazards can be addressed before they manifest on-site. This holistic, team-based approach demonstrates that health and safety outcomes are heavily influenced by early project decisions and collective stakeholder engagement rather than solely by site-level measures or individual worker behaviour.

In conclusion, the successful application of buildability to enhance health and safety depends on the active participation and collaboration of all project stakeholders. Designers, contractors, clients, and safety professionals each contribute critical knowledge and influence that, when combined, create safer construction processes, improve constructability, and embed safety as a fundamental aspect of project delivery. Effective stakeholder collaboration ensures that buildability-driven safety is not an isolated concept but a practical and integral part of construction management.

6.2.7 Barriers to Integrating Buildability and Safety

Despite the well-established conceptual link between buildability and health and safety, several barriers continue to hinder their effective integration in construction projects. These barriers are multifaceted, encompassing organisational structures, professional expertise, procurement methods, and project management practices. Understanding these challenges is critical for developing strategies to embed buildability-driven safety more effectively across the project lifecycle.

One of the primary barriers is the fragmentation of project teams. Construction projects typically involve multiple stakeholders, including clients, designers, contractors, subcontractors, and suppliers, often operating under separate contractual arrangements. This fragmentation can lead to poor communication, unclear allocation of responsibilities, and limited collaboration during critical design and planning phases. When information regarding site hazards, sequencing, or practical construction constraints is not shared effectively, opportunities to design out risks or optimise buildability are frequently missed. Consequently, latent hazards persist into the construction phase, increasing reliance on procedural controls and worker compliance rather than systemic risk reduction.

Another significant barrier is limited designer awareness and understanding of construction risks. Many designers, particularly architects and engineers, receive little formal education in construction methods, site hazards, or OHS principles. As a result, they may struggle to anticipate how design decisions, such as complex geometries, inadequate access provisions, or congested layouts, translate into on-site hazards. Insufficient exposure to practical construction experience reduces their ability to integrate buildability considerations effectively, leaving critical safety opportunities unaddressed. This knowledge gap is compounded by a lack of formal training in buildability principles and proactive safety design approaches, including PtD and DfS.

Traditional procurement methods also pose barriers to integrating buildability and safety. For example, the widely used design–bid–build model typically separates the design and construction phases, limiting early contractor involvement and reducing the ability to incorporate practical construction insights into design. This separation constrains collaborative decision-making and may result in designs that are challenging to construct safely. Even in design–build or integrated project delivery models, tight schedules and budget pressures often limit the time available for thorough evaluation of safety implications and constructability considerations. Time constraints during design stages can lead to superficial attention to buildability and safety, with designers prioritising aesthetics, functionality, and cost over practical constructability and risk mitigation.

Addressing these barriers requires a combination of organisational, educational, and regulatory interventions. Organisational change is needed to promote integrated, multidisciplinary collaboration early in project planning. Enhanced training and professional development for designers can improve awareness of construction hazards and buildability principles. Regulatory frameworks, such as the UK's Construction (Design and Management) Regulations, play a critical role by formalising designer responsibilities for health and safety, reinforcing the expectation that risk management begins during design rather than on-site.

In summary, fragmentation, limited designer expertise, restrictive procurement practices, and time pressures are key barriers to integrating buildability and safety. Overcoming these challenges necessitates systemic change, including collaboration-focused organisational structures, targeted education and training, and supportive regulations that embed safety and buildability as fundamental components of project delivery.

6.2.8 Empirical Evidence Supporting the Link

A growing body of empirical research provides strong support for the conceptual link between buildability and improved health and safety outcomes in construction projects. These studies consistently demonstrate that design decisions directly influence the risk exposure of workers and that adopting buildability-focused strategies can significantly reduce accidents, injuries, and occupational health hazards. By examining real-world projects, researchers have highlighted practical approaches through which buildability translates into measurable safety improvements.

Gambatese et al. (2008) conducted a comprehensive study identifying design strategies that mitigate construction risks by embedding safety considerations into the design process. Key strategies include increasing component repetition, simplifying complex details, providing adequate access and egress for construction activities, and ensuring clear sequencing of work. By implementing these approaches, designers can reduce the likelihood of unsafe workarounds, minimise manual handling, and limit work at height. Projects that integrate such buildability principles have consistently demonstrated lower accident rates and improved overall safety performance, highlighting the tangible benefits of proactive, design-led safety management. These findings underscore that safety is not solely dependent on-site–level interventions but is strongly influenced by upstream decisions made during design and planning.

Prefabrication and modular construction represent particularly compelling examples of the health and safety benefits of improved buildability. Research by Gibb and Isack (2003) indicates that shifting construction activities to controlled factory environments significantly reduces exposure to common site hazards, including falls, struck-by incidents, and environmental risks such as extreme weather. Prefabricated components can be manufactured under safe, regulated conditions and then assembled on-site using streamlined processes, reducing the duration and complexity of high-risk activities. Modular construction further allows for off-site assembly of entire building sections, limiting the number of workers exposed to hazards and enhancing predictability in safety performance.

In addition, empirical studies highlight that buildability-driven design interventions not only prevent accidents but also positively influence organisational and cultural aspects of safety. Projects that prioritise constructability often exhibit improved coordination between designers, contractors, and safety professionals, promoting proactive risk identification and mitigation. This integrated approach aligns with the principles of PtD and DfS, demonstrating that buildability serves as a mechanism for embedding systemic safety improvements rather than relying solely on procedural controls or PPE.

Furthermore, longitudinal analyses of construction projects reveal that the benefits of buildability extend beyond immediate construction phases. Simplified designs, standardised components, and accessible layouts contribute to safer maintenance and operational practices, ensuring that health and safety outcomes are sustained throughout the building's lifecycle. These findings reinforce the notion that buildability is not merely a productivity or cost-related concern but a critical determinant of both worker safety and long-term operational well-being.

In summary, empirical evidence robustly supports the link between buildability and improved health and safety performance. Design strategies that prioritise constructability, prefabrication, and modularity reduce site hazards, improve workflow predictability, and foster proactive safety management. These findings provide compelling justification for integrating buildability principles into design, planning, and regulatory frameworks to achieve safer, more efficient, and sustainable construction outcomes.

In conclusion, the conceptual link between buildability and health and safety is grounded in the understanding that design decisions fundamentally shape construction risk. Buildability provides a practical framework for integrating safety considerations into design, enabling hazards to be eliminated or reduced at their source. Supported by regulatory developments, safety theories, and empirical evidence, buildability transforms health and safety from a reactive site-based function into a proactive, systemic process. Recognising and strengthening this link is essential for achieving sustainable improvements in construction health and safety performance and sets the foundation for examining specific buildability factors influencing safety outcomes, as explored in the subsequent section.

6.3 Key Buildability Factors Influencing Health and Safety Outcomes

Buildability directly influences health and safety outcomes through specific design and planning decisions that shape how construction activities are executed. While the conceptual relationship between buildability and health and safety establishes the importance of early design intervention, it is the application of concrete buildability factors that translates this relationship into measurable safety improvements. These factors determine construction complexity, work methods, site conditions, and worker exposure to hazards. This section critically examines the key buildability factors that most significantly affect health and safety outcomes in construction projects.

6.3.1 Design Simplicity and Standardisation

Design simplicity is widely recognised as one of the most influential buildability factors impacting construction health and safety. Complex designs, characterised by irregular geometries, bespoke components, and intricate detailing, often necessitate specialised construction methods, extensive temporary works, and high levels of manual handling. These factors inherently increase the likelihood of accidents, as workers are exposed to elevated risks associated with working at height, handling heavy or awkward components, and operating in congested or constrained spaces. Additionally, complexity in design frequently leads to misunderstandings or misinterpretations of drawings and specifications on-site, resulting in rework, rushed activities, and unsafe improvisations, which cumulatively escalate hazard exposure and accident probability.

Simplicity in design, by contrast, facilitates safer and more predictable construction processes. Simplified designs reduce the number of unique components, standardise construction methods, and minimise the need for temporary structures or complex sequencing. This clarity enables workers to understand tasks more readily, anticipate potential risks, and follow consistent procedures, thereby lowering the likelihood of errors and accidents. By eliminating unnecessary intricacy, simple designs also support better supervision and oversight, allowing site managers to more effectively monitor compliance with safety procedures and identify potential hazards before they result in incidents.

Closely linked to simplicity is the principle of standardisation, which enhances safety by promoting repetition and familiarity in construction tasks. When components are standardised, workers repeatedly perform similar installations using consistent methods. This repetition fosters skill development, reduces cognitive load, and embeds safe working practices into routine operations. For example, standardised structural elements or façade panels allow predictable planning of lifting operations, use of access systems, and sequencing of work, thereby reducing the uncertainty that often contributes to unsafe behaviours. Standardisation also supports the use of prefabricated or modular elements, further decreasing exposure to high-risk activities such as working at height or in confined spaces.

From a cognitive perspective, both simplicity and standardisation reduce the mental and physical demands placed on workers and supervisors. Complex or unique construction tasks require greater attention, decision-making under pressure, and problem-solving in real time, all of which can increase stress, fatigue, and the potential for mistakes. Simple and standardised designs minimise these demands, enabling workers to focus on executing tasks safely rather than improvising solutions to unforeseen challenges. Supervisors benefit similarly, as monitoring fewer unique tasks makes it easier to ensure adherence to safe practices and detect hazards promptly.

In summary, design simplicity and standardisation are fundamental to improving construction health and safety. By reducing design complexity, promoting repetition, and enabling predictable workflows, these buildability factors minimise hazard exposure, support safer construction methods, and foster a culture of consistent safe practice. Integrating simplicity and standardisation into the design process not only enhances worker safety but also contributes to overall project efficiency and quality, demonstrating that buildability and health and safety are mutually reinforcing objectives.

6.3.2 Prefabrication, Pre-assembly, and Modular Construction

Prefabrication, pre-assembly, and modular construction represent some of the most impactful strategies for enhancing buildability and improving health and safety outcomes in construction projects. These approaches involve the off-site manufacture of building components or entire modules in controlled factory environments, followed by transportation and on-site installation. By relocating high-risk activities away from the often hazardous and unpredictable construction site, prefabrication significantly reduces worker exposure to common site hazards such as working at height, manual handling of heavy materials, adverse weather conditions, and congested work areas.

Factory environments provide several advantages over traditional on-site construction in terms of safety. They typically offer better ergonomics, allowing workers to perform tasks in comfortable and controlled postures, with reduced strain from lifting, bending, or repetitive handling. Supervision and quality control are also more consistent, with standardised procedures and routine checks ensuring that tasks are performed safely and accurately. These controlled conditions reduce the likelihood of human error and mitigate occupational health risks associated with prolonged exposure to dust, vibration, noise, and hazardous substances. Collectively, these factors contribute to lower accident rates and improved long-term health outcomes for the workforce.

On-site installation of prefabricated or modular components further enhances safety by simplifying construction processes. Prefabricated elements are often designed for rapid assembly, requiring fewer workers and less time at the site, which reduces both the duration and intensity of exposure to residual hazards. Additionally, standardisation and repetition in assembling prefabricated components facilitate safer, more predictable workflows. Empirical research supports these benefits, with studies demonstrating that projects employing higher levels of prefabrication consistently achieve lower accident rates and improved overall safety performance compared to conventional construction methods. The reduction in complex on-site construction tasks also decreases cognitive load for workers and supervisors, contributing to more reliable compliance with safe working practices.

Despite these advantages, prefabrication and modular construction are not without their own risks. Transportation of large modules and lifting operations on-site introduce new hazards, such as the potential for dropped components, collisions, or unstable positioning during installation. These risks highlight the need for holistic buildability planning that addresses the entire lifecycle of prefabricated elements, from design through delivery and erection. Effective mitigation strategies include careful component sizing, the incorporation of designed lifting points, detailed sequencing plans, and thorough risk assessments for transportation and installation activities. By considering these factors during design and project planning, prefabrication can maximise its safety benefits while minimising new hazards.

In summary, prefabrication, pre-assembly, and modular construction exemplify the practical application of buildability principles to enhance construction safety. By shifting high-risk work to controlled environments, reducing on-site exposure, and enabling efficient, standardised assembly, these approaches significantly improve health and safety performance. When complemented by careful planning for transportation and lifting, prefabrication demonstrates that buildability is a powerful tool for systemic risk reduction and safer, more efficient construction processes.

6.3.3 Construction Sequencing and Work Planning

Construction sequencing is a pivotal buildability factor that has a direct and profound impact on health and safety outcomes on construction sites. Sequencing refers to the logical order in which construction activities are planned and executed, encompassing the coordination of trades, installation of components, and delivery of materials. Poorly planned sequences can create overlapping work activities, congested work areas, and conflicting operations, all of which substantially elevate the risk of accidents. For example, when structural, mechanical, and finishing works occur simultaneously in confined spaces, the likelihood of incidents, such as falls, struck-by accidents, and equipment collisions, increases significantly. Effective sequencing, therefore, is essential not only for efficiency but also for minimising exposure to hazards.

Designers play a crucial role in determining safe and efficient construction sequences through their decisions regarding structural systems, component connections, and assembly logic. Designs that incorporate clear, logical, and linear sequences enable construction to progress in an orderly and controlled manner, reducing the need for temporary works, improvisation, or rework, all of which are common contributors to accidents. For instance, predefining the sequence of structural assembly, followed by mechanical, electrical, and plumbing installations and finishing trades, can prevent interference between teams and reduce congestion in critical areas of the site. Additionally, well-planned sequences facilitate better site layout management, allowing safe movement of personnel, equipment, and materials.

Early consideration of sequencing during the design phase also supports safer planning of lifting operations and material deliveries. Cranes and lifting equipment are frequently involved in high-risk activities, particularly when multiple trades compete for space or when components must be lifted over partially completed structures. By aligning the design and sequencing logic, designers can ensure that lifting paths, storage areas, and delivery schedules are optimised to minimise potential hazards. This proactive approach reduces the likelihood of dropped objects, collisions, and other lifting-related incidents, which remain among the most common causes of construction injuries.

Integrating sequencing into design also enhances proactive safety management by enabling the identification of high-risk activities well before work begins on-site. Early identification allows for hazard elimination or reduction through design decisions, such as simplifying component assembly, standardising connection methods, or incorporating prefabricated elements that reduce on-site complexity. Sequencing consideration thus reinforces the principle that safety is not solely a function of site supervision or worker behaviour but is fundamentally influenced by upstream design decisions.

In summary, construction sequencing and work planning are critical buildability factors that directly influence health and safety outcomes. Logical, well-considered sequencing reduces trade overlap, minimises congestion, improves lifting and material handling operations, and mitigates exposure to hazards. When integrated into design and project planning, sequencing supports proactive risk management, emphasises hazard elimination, and transforms construction safety from a reactive site-based function into a systematic, design-led process. By prioritising sequencing as a core element of buildability, projects can achieve both safer working conditions and more efficient construction outcomes.

6.3.4 Access, Egress, and Site Logistics

Safe access and egress are fundamental components of construction health and safety, particularly given that falls from height remain the leading cause of fatalities in the sector worldwide. Ensuring that workers, materials, and equipment can move safely across the site requires careful planning, which is most effective when incorporated at the design stage. Buildability-focused design provides the opportunity to embed safe access and egress routes from the outset, rather than treating them as temporary or improvised measures during construction. By considering these elements early, designers can prevent common hazards and reduce reliance on reactive safety interventions.

Design decisions related to floor layouts, structural openings, stair locations, and façade systems significantly affect the practicality of implementing safe access systems. For example, sufficient space for scaffolding, guardrails, edge protection, and temporary staircases must be incorporated into the building's geometry and assembly sequence. Gambatese et al. (2008) emphasise that when these elements are neglected or inadequately planned, contractors are often forced to improvise, creating unsafe workarounds that increase the risk of falls, trips, and manual handling injuries. Well-considered access not only safeguards workers during construction but also facilitates safer maintenance and operational activities over the building's lifecycle, highlighting the broader impact of buildability on long-term occupational health.

Closely intertwined with access and egress is site logistics, encompassing material storage, waste management, vehicle circulation, and pedestrian pathways. Poorly planned logistics can result in congested sites, blocked access, and unsafe interactions between workers, vehicles, and machinery, contributing to slips, trips, collisions, and struck-by incidents. By integrating site logistics considerations into the design process, buildability principles ensure that spatial constraints are addressed, material flows are optimised, and critical circulation routes remain unobstructed. For instance, designing dedicated laydown areas, clear delivery paths, and safe material handling zones reduces congestion, minimises exposure to hazards, and enhances efficiency.

Moreover, integrating access, egress, and logistics considerations during design supports proactive safety management by anticipating potential hazards and planning to eliminate them before work commences on-site. Early collaboration between designers, contractors, and safety professionals allows the identification of practical constraints, ensuring that access routes and site layouts are feasible and compliant with safety standards. This proactive planning reduces the need for temporary, high-risk solutions and improves predictability in construction operations.

From a cognitive and operational perspective, well-planned access, egress, and logistics reduce worker stress and confusion, enabling personnel to focus on tasks safely rather than navigating congested or unsafe conditions. It also facilitates supervision and monitoring, allowing site managers to enforce safety practices more effectively.

In summary, access, egress, and site logistics are critical determinants of construction safety that are strongly influenced by buildability-focused design. By embedding safe movement routes, considering structural layouts, and planning efficient material flows, designers can minimise fall hazards, reduce congestion, and create safer, more organised construction environments. Integrating these considerations early reinforces the systemic approach of buildability, demonstrating that health and safety outcomes are significantly shaped by upstream design and planning decisions.

6.3.5 Structural Systems and Temporary Works

The selection of structural systems is a critical factor influencing both buildability and health and safety in construction projects. Structural systems not only determine how a building will perform in its completed state but also significantly affect the safety and practicality of its assembly. Systems that require extensive temporary works, such as formwork, shoring, scaffolding, and propping, introduce additional hazards that must be managed throughout erection and dismantling. These hazards include risks of falls, struck-by incidents, instability, and even partial or total collapse of temporary structures, all of which pose significant threats to workers. Consequently, integrating buildability considerations into structural design is essential to minimise reliance on hazardous temporary solutions and to create inherently safer construction processes.

In-situ concrete systems, while versatile and commonly used, often demand complex formwork, propping, and shoring arrangements to support structural elements during curing. Each layer of temporary works adds complexity and exposure to risk, requiring meticulous planning, supervision, and skilled labour to ensure safety. Any error or oversight in design or erection can have severe consequences, as temporary structures are frequently cited as critical contributors to accidents. By contrast, structural systems such as steel framing or precast concrete elements offer advantages in buildability and safety. These systems generally allow for faster assembly, reduced reliance on extensive formwork, and fewer high-risk manual handling operations. For example, precast components can be manufactured under controlled factory conditions and installed using predictable lifting procedures, thereby reducing time spent on-site and minimising exposure to hazards.

However, even systems with inherent buildability benefits require careful attention to stability and connection design during erection. Improper sequencing, poorly designed temporary supports, or inadequately braced components can lead to instability and collapse, demonstrating that safety risks remain present even with "easier-to-build" systems. Integrating temporary and permanent stability considerations into early structural design is therefore crucial. By anticipating erection sequences, designing for self-supporting stages where possible, and incorporating clear specifications for temporary works, designers can significantly reduce hazards associated with critical construction phases.

Temporary works design is frequently identified as a weak point in construction safety management, often addressed only once on-site constraints are fully understood. This reactive approach can result in improvised or last-minute solutions that compromise both buildability and safety. Incorporating temporary works planning into the early design stage ensures that all necessary supports, access provisions, and sequencing are considered systematically. Such proactive design not only enhances worker safety but also contributes to efficiency by reducing delays, rework, and site congestion associated with complex temporary arrangements.

In summary, structural systems and temporary works are closely intertwined with buildability and health and safety outcomes. Selecting systems that minimise the need for temporary works, designing components for inherent stability, and integrating temporary works considerations into early design stages all contribute to safer construction environments. By embedding these principles, designers can reduce on-site hazards, enhance constructability, and ensure that safety is an integral component of both structural performance and construction methodology.

6.3.6 Materials Selection and Handling

Material selection is a critical factor in construction that directly impacts both buildability and health and safety outcomes. The types of materials used on a project influence manual handling requirements, exposure to hazardous substances, and the potential for long-term occupational health issues. Heavy, awkward, or irregularly shaped materials increase the likelihood of musculoskeletal injuries, including strains, sprains, and chronic back problems. Simultaneously, materials containing hazardous substances, such as asbestos, silica, lead-based products, or certain chemical coatings, pose significant risks through inhalation, skin contact, or prolonged exposure, potentially leading to respiratory illnesses, dermatological conditions, or other chronic diseases. Consequently, thoughtful material selection is essential not only for construction efficiency but also for safeguarding worker health.

A buildability-oriented approach to material selection prioritises components that are easier to handle, transport, and install. Lightweight and modular materials, for instance, reduce physical strain on workers and minimise the need for lifting equipment or complex rigging arrangements. Engineered or composite products are often designed with standardised dimensions and tolerances, enabling safer and more predictable installation processes. Prefabricated panels, modular units, and pre-assembled systems exemplify how material choices can directly reduce on-site risks by limiting manual handling, reducing exposure to heights, and streamlining construction sequencing. By selecting materials that are compatible with efficient and straightforward construction methods, designers and project managers can embed safety into the fabric of the building process, rather than relying solely on-site–based controls.

Beyond immediate accident prevention, material selection plays a critical role in protecting long-term occupational health. Specifying low-dust, low-toxicity, or chemically inert materials reduces inhalation and dermal exposure for workers, supporting broader health objectives. For example, substituting traditional cement with blended cements or using non-toxic adhesives and finishes can significantly lower the risk of respiratory or skin conditions. Similarly, materials that are resistant to degradation or that require minimal maintenance reduce future occupational exposure risks during repair or refurbishment activities. By considering both short-term and long-term implications, material selection reinforces the principle that health and safety management extends beyond immediate construction hazards to encompass holistic worker well-being.

Material handling practices are equally influenced by buildability considerations. Standardised component sizes, modular units, and well-defined lifting points facilitate safer manual handling, reduce the likelihood of improper lifting techniques, and decrease reliance on temporary rigging or ad-hoc handling solutions. Training workers in the safe handling of selected materials further enhances safety outcomes and ensures that buildability decisions translate into practical on-site benefits.

In summary, material selection is a vital intersection between buildability and health and safety. By prioritising lightweight, modular, and low-risk materials, designers and contractors can minimise manual handling demands, reduce exposure to hazardous substances, and safeguard long-term worker health. This approach demonstrates that buildability is not solely a matter of construction efficiency but a powerful mechanism for embedding systemic health and safety benefits throughout the project lifecycle, extending protection from immediate construction activities to long-term occupational well-being.

6.3.7 Repetition, Learning, and Workforce Familiarity

Repetition is a fundamental buildability factor that has a profound influence on construction health and safety. By creating predictable and standardised workflows, repetition enables workers to gain proficiency in specific tasks, while allowing supervisors to monitor performance more effectively. Repetition reduces variability in work methods, which is widely recognised as a key contributor to accidents on construction sites. When workers repeatedly perform similar operations using consistent techniques, they develop muscle memory, anticipate potential hazards, and internalise safe work practices. This familiarity translates into fewer errors, reduced improvisation, and safer overall performance.

Standardisation of components and construction methods reinforces the benefits of repetition. For example, using prefabricated or modular elements allows tasks to be executed in a consistent manner across multiple locations or phases of a project. Standardised work facilitates the development of detailed method statements, comprehensive risk assessments, and clear operating procedures. These documents serve as essential tools for hazard identification and control, providing workers with step-by-step guidance that reduces uncertainty and enhances compliance with safety protocols. Furthermore, supervisors benefit from repetition because predictable workflows make it easier to identify deviations, intervene proactively, and reinforce safe practices before incidents occur.

Conversely, projects characterised by bespoke designs, complex geometries, or constantly changing requirements limit opportunities for repetition. Each task may differ in scope, methods, or constraints, forcing workers to rely heavily on individual judgment and improvisation. This variability increases the likelihood of errors, miscommunication, and unsafe shortcuts, all of which elevate accident risk. Non-repetitive tasks also reduce the effectiveness of standard training programs, as each activity may require unique skills or knowledge. In such contexts, the benefits of accumulated learning and familiarity are diminished, making safety management more challenging and reliant on reactive measures.

Repetition also facilitates continuous improvement and proactive interventions. When tasks are repeated systematically, organisations can monitor performance, identify recurring hazards, and implement incremental improvements to work methods. Targeted training programs can be developed based on observed patterns, reinforcing safe practices and addressing specific risks associated with repetitive tasks. Over time, this creates a culture of learning and adaptation, where both workers and supervisors contribute to ongoing enhancement of safety performance.

From a broader perspective, repetition illustrates the close link between buildability and health and safety. Designs that promote repeatable, predictable construction sequences not only improve efficiency but also inherently support safer working conditions. By reducing variability, fostering familiarity, and enabling learning, repetitive workflows allow safety to be embedded within the construction process rather than treated as a reactive or supplementary activity.

In summary, repetition, learning, and workforce familiarity are integral to buildability-driven safety. Standardised and predictable tasks enable skill development, hazard anticipation, and proactive supervision, while reducing variability and reliance on individual judgment. Projects that embrace repetition and standardisation enhance worker competence, support continuous improvement, and achieve measurable improvements in health and safety performance, demonstrating that buildability is a key driver of safer construction environments.

6.3.8 Maintainability and Whole-Life Safety

Buildability is not limited to facilitating efficient and safe construction; it also plays a critical role in maintaining long-term OHS throughout a building's lifecycle. Designs that focus solely on immediate constructability often neglect considerations related to maintenance, repair, and eventual demolition. This oversight can create substantial hazards for facility managers, maintenance personnel, and contractors during the operational life of the asset. Common risks include working at height without safe access, handling heavy or awkward components in confined spaces, and performing repetitive or strenuous tasks under unsafe conditions. By integrating maintainability into design decisions, buildability principles extend safety benefits far beyond the construction phase.

One of the primary ways to achieve maintainability and whole-life safety is through the inclusion of permanent access systems. These may include fixed staircases, ladders, catwalks, walkways, and guardrails that provide safe routes for inspection, repair, and routine maintenance activities. Permanent access reduces reliance on temporary structures such as scaffolding or mobile platforms, which are inherently higher risk due to their transient nature and potential for improper installation. By planning access solutions early, designers can ensure that maintenance tasks can be performed safely, efficiently, and with minimal disruption to operations.

Safe lifting points and handling provisions are also critical components of maintainability. Buildings and industrial facilities often require the movement or replacement of heavy equipment, machinery, or structural components. Incorporating clearly defined lifting points, appropriate anchorage, and designated handling zones into the design allows maintenance personnel to execute these tasks with reduced exposure to falls, strains, or collisions. When these considerations are integrated into the early design stages, the risks associated with future repair or replacement work are significantly mitigated, reflecting the systemic nature of buildability as a safety enabler.

Maintaining clear maintenance zones further enhances safety by ensuring that work areas are free from obstructions and that hazardous interactions with operational processes or other personnel are minimised. Well-planned spatial arrangements facilitate safer movement of workers and materials, reduce congestion, and provide clear delineation between operational and maintenance activities. These design decisions directly influence both efficiency and safety, demonstrating that buildability and long-term health and safety are inherently linked.

The whole-life perspective reinforces the argument that buildability is integral to sustainable health and safety performance. By extending considerations beyond construction to include operation, maintenance, and eventual decommissioning, designers can embed safety into the entire lifecycle of an asset. This approach not only protects workers from immediate risks but also safeguards their health and well-being over decades of service, reducing the likelihood of occupational injuries and chronic conditions associated with poor design.

In summary, maintainability and whole-life safety are key dimensions of buildability that extend the benefits of proactive design far beyond construction. Incorporating permanent access systems, safe lifting points, and designated maintenance zones ensures that future operations can be conducted safely, efficiently, and predictably. This lifecycle-focused approach demonstrates that buildability is not merely a tool for construction efficiency but a critical enabler of sustainable, long-term health and safety outcomes.

6.3.9 Digital Tools and Buildability Analysis

The integration of digital technologies into construction design and planning has significantly advanced the assessment of buildability and health and safety outcomes. Among these technologies, BIM has emerged as a particularly transformative tool, enabling designers, contractors, and safety professionals to visualise, analyse, and optimise construction processes before work begins on-site. By creating a comprehensive digital representation of the built environment, including structural, mechanical, electrical, and architectural elements, BIM facilitates detailed scrutiny of construction sequences, identification of potential conflicts, and simulation of site conditions under various scenarios. This capability enhances both buildability and proactive safety management.

BIM and other digital tools enable early identification of high-risk activities, allowing designers and contractors to explore alternative solutions to minimise hazards. For example, through 3D visualisation and sequencing simulations, stakeholders can detect areas where trades may conflict, assess the feasibility of safe access routes, and determine whether lifting operations or material handling could pose risks. Clash detection tools can highlight spatial conflicts, such as overlapping ductwork or congested scaffolding zones, which could otherwise result in unsafe workarounds or rework during construction. By resolving these issues during design rather than on-site, the potential for accidents is substantially reduced, demonstrating the powerful link between digital buildability analysis and health and safety performance.

Digital tools also support optimisation of construction sequencing and logistics. Simulating construction processes within a virtual environment enables stakeholders to evaluate the impact of different sequencing strategies, identify bottlenecks, and plan safe material flows. For instance, crane lift simulations and virtual walkthroughs can determine safe access paths and verify that prefabricated components can be installed without exposing workers to unnecessary hazards. Such predictive capabilities allow for proactive hazard elimination, shifting safety management upstream and reinforcing the principle that risks are largely "designed in" or "designed out" during early project stages.

Furthermore, BIM facilitates collaboration and information sharing among project stakeholders, which is essential for buildability-driven safety. Designers, contractors, clients, and safety professionals can work within a shared digital model, improving communication and enabling informed decision-making. This collaborative environment ensures that buildability considerations, including structural simplicity, temporary works requirements, access, and material handling, are consistently evaluated and integrated across the project lifecycle. By centralising information and providing a clear visual context, digital tools enhance the ability to embed systemic safety measures rather than relying solely on-site–level interventions.

While digital technologies do not automatically guarantee improved safety, they provide a robust framework for embedding buildability-led safety into design processes. By enabling early hazard identification, scenario analysis, and collaborative problem-solving, tools such as BIM translate design intent into safer construction practices and more predictable on-site outcomes.

In summary, digital tools and buildability analysis represent a critical evolution in construction planning. Through simulation, visualisation, and predictive modelling, they allow proactive identification and mitigation of hazards, optimisation of construction sequences, and improved stakeholder collaboration. Integrating these technologies into design enhances both buildability and health and safety, demonstrating the potential of digital solutions to transform construction practices from reactive risk management to a systematic, design-led approach.

Key buildability factors, including design simplicity, prefabrication, sequencing, access planning, structural systems, material selection, and maintainability, have a profound influence on health and safety outcomes in construction. These factors determine the level of risk workers are exposed to and the effectiveness of safety controls. By addressing safety at its source through informed design decisions, buildability transforms health and safety from a reactive site-based function into a proactive, integrated component of project delivery. Recognising and applying these buildability factors is essential for achieving sustained improvements in construction health and safety performance.

6.4 Chapter Conclusion

This chapter has demonstrated that buildability and health and safety are intrinsically interconnected, with design decisions serving as a critical determinant of construction risk. In an industry historically characterised by high accident rates, complex site conditions, and fragmented project delivery, improving safety performance requires more than compliance with regulations; it demands a proactive, design-led approach. By positioning buildability as a systemic risk control mechanism, this chapter has shown that safer construction outcomes can be achieved when constructability principles are embedded early and consistently throughout the project lifecycle.

The overview of construction health and safety highlighted the unique nature of construction work, including its temporary worksites, dynamic environments, and diverse workforce. Traditional safety approaches have often relied heavily on administrative controls, PPE, and reactive measures following incidents. While regulatory frameworks and legal responsibilities have strengthened accountability, design-related factors remain a significant contributor to construction accidents. Recognising this, the industry has gradually shifted toward prevention-focused strategies, emphasising safety culture, occupational health awareness, and early risk mitigation.

Central to this shift is the conceptual link between buildability and safety. The evolution of buildability thinking aligns closely with the principles of PtD and DfS, both of which emphasise eliminating hazards at their source. When design teams incorporate construction knowledge, anticipate site constraints, and simplify complex elements, they inherently reduce exposure to risk. Buildability thus functions as a proactive mechanism that integrates safety considerations into decision-making processes, rather than treating safety as an add-on during the construction phase.

This chapter further identified key buildability factors that directly influence health and safety outcomes. Design simplicity and standardisation reduce uncertainty and variability on-site, lowering the likelihood of errors and unsafe practices. Prefabrication, pre-assembly, and modular construction minimise hazardous on-site activities by transferring work to controlled environments. Effective construction sequencing, careful planning of access and logistics, and thoughtful consideration of structural systems and temporary works significantly mitigate operational risks. Similarly, informed materials selection, repetition that enhances workforce familiarity, and attention to maintainability extend safety benefits beyond construction into the operational phase of assets. The integration of digital tools, such as BIM-based risk visualisation and buildability analysis, further strengthens the ability to anticipate, assess, and manage hazards.

Importantly, this chapter acknowledged the barriers to integrating buildability and safety, including fragmented project structures, limited early contractor involvement, cost-driven decision-making, and insufficient stakeholder collaboration. Overcoming these challenges requires cultural change, leadership commitment, and interdisciplinary coordination. Empirical evidence increasingly supports the conclusion that projects incorporating buildability principles experience improved safety performance, reduced rework, and enhanced overall efficiency.

In conclusion, buildability should be recognised not only as a driver of cost and productivity improvements but also as a foundational pillar of construction health and safety. By embedding constructability considerations into design, planning, and execution processes, stakeholders can move from reactive compliance to proactive risk prevention. Ultimately, integrating buildability and safety fosters safer workplaces, protects worker well-being, and contributes to the long-term sustainability and resilience of the construction industry.

Bibliography

Ahmed, F., Gkantou, M., Nikitas, G., Ferentinou, M., Bras, A., & Riley, M. (2025). A review of recent developments in structural elements of modular steel building systems. *Journal of Building Engineering*, 111, 113087. https://doi.org/10.1016/j.jobe.2025.113087

Alashwal, A., & Fong P.S. (2015). Empirical study to determine fragmentation of construction projects. *Journal of Construction Engineering and Management*, 141(7), 04015016. https://doi.org/10.1061/(ASCE)CO.1943-7862.0000986

Al Hamdani, M.S., & Alsaad, L.N. (2024). Occupational lung diseases among workers in the construction industry in the MENA region: A literature review. *General Medicine*, 26(8), 1823–1832.

Ali, A.M., & Nouban, F. (2024). Analyzing the role of leadership in promoting a positive safety culture in the construction sector. *World Journal of Advanced Research and Reviews*, 21, 882–887. https://doi.org/10.30574/wjarr.2024.21.1.0067

Almaskati, D., Kermanshachi, S., Pamidimukkala, A., Loganathan, K., & Yin, Z. (2024). A review on construction safety: Hazards, mitigation strategies, and impacted sectors. *Buildings*, 14(2), 526. https://doi.org/10.3390/buildings14020526

Al-Naser, N.K., & Al-Tabtabai, H. (2025). The impact of safety violations on construction project performance: A case study of the ADFA project. *Journal of Engineering Research*, 13(4), 3037–3050. https://doi.org/10.1016/j.jer.2024.12.017

Andi, & Minato, T. (2003). Design documents quality in the Japanese construction industry: factors influencing and impacts on construction process. *International Journal of Project Management*, 21(7), 537–546. https://doi.org/10.1016/S0263-7863(02)00083-2

Ang, Y.Q., Chew, L.W., & Samuelson, H. (2025). Healthy building design strategies: A cross-topic systematic review. *Journal of Building Engineering*, 105, 112421. https://doi.org/10.1016/j.jobe.2025.112421

Anim Manu, B. (2024). Innovative construction materials: Advancing sustainability, durability, efficiency, and cost-effectiveness in modern infrastructure. *International Journal of Research Publication and Reviews*, 5(12), 4987–4999. https://doi.org/10.55248/gengpi.5.1224.0215

Anzagira, L.F., Avogo, J.A., & Tengan, C. (2025). Impact of work safety culture on safety measures in construction firms: A systematic review. *Scientific Journal of Engineering, and Technology*, 2(2), 1–8. https://doi.org/10.69739/sjet.v2i2.666

Araújo, L.L.F., Saldanha, M.C.W., Gohr, C.F., & Nascimento, G.H.P. (2022). Improvement factors of constructability and occupational safety on project life cycle phases. *Automation in Construction*, 138, 104227. https://doi.org/10.1016/j.autcon.2022.104227

Arif, M., Blismas, N., & Wakefield, R. (2009). Drivers, constraints and the future of offsite manufacture in Australia. *Construction Innovation: Information Process Management*, 9(1), 72–83. https://doi.org/10.1108/14714170910931552

Bajwa, A.U.R., Siriwardana, C., Shahzad, W., & Naeem, M.A. (2025). Material selection in the construction industry: A systematic literature review on multi-criteria decision making. *Environment Systems and Decisions*, 45(8). https://doi.org/10.1007/s10669-025-10001-w

Bataglin, F., Viana, D.D., & Formoso, C.T. (2022). Design principles and prescriptions for planning and controlling engineer-to-order industrialized building systems. *Sustainability*, 14(24), 16822. https://doi.org/10.3390/su142416822

Bautista-Bernal, I., Quintana-García, C., & Marchante-Lara, M. (2024). Safety culture, safety performance and financial performance. A longitudinal study. *Safety Science*, 172, 106409. https://doi.org/10.1016/j.ssci.2023.106409

Behm, M. (2005). Linking construction fatalities to the design for construction safety concept. *Safety Science*, 43(8), 589–611. https://doi.org/10.1016/j.ssci.2005.04.002

Bilau, A.A., Ajagbe, M.A., Sholanke, A.B., & Sani, T.A. (2015). Impact of employee turnover in small and medium construction firms: A literature review. *International Journal of Engineering Research & Technology*, 4(2), 976–984.

Chetty, D.R.V., Boojhawon, R., Bhagwant, S., & Levy, L. (2024). Factors affecting the occupational safety and health of small and medium enterprises in the construction sector of Mauritius. *Social Sciences & Humanities Open*, 10, 100964. https://doi.org/10.1016/j.ssaho.2024.100964

CIRIA. (1983). Buildability: An assessment. *CIRIA Publications*, Special Report, London, p. 26.

Cooper, M.D. (2000). Towards a model of safety culture. *Safety Science*, 36(2), 111–136. https://doi.org/10.1016/S0925-7535(00)00035-7

Daniel, E.I., Oshodi, O.S., Nwankwo, N.I., Emuze, F.A., & Chinyio, E. (2025). The use of digital technologies in construction safety: A systematic review. *Buildings*, 15(8), 1386. https://doi.org/10.3390/buildings1508138

Fayyaz, K., Shahzaib, M., Aziz, A., Irfan, M., Salah Alaloul, W., & Musarat, M.A. (2025). Cultural factors impacting health and safety (H&S) practices in a developing construction economy. *Sustainability*, 17(3), 911. https://doi.org/10.3390/su17030911

Firoozi, A.A., Firoozi, A.A., Oyejobi, D.O., Avudaiappan, S., Flores, E.S. (2024). Emerging trends in sustainable building materials: Technological innovations, enhanced performance, and future directions. *Results in Engineering*, 24, 103521. https://doi.org/10.1016/j.rineng.2024.103521

Gambatese, J.A., Behm, M., & Rajendran, S. (2008). Design's role in construction accident causality and prevention: Perspectives from an expert panel. *Safety Science*, 46(4), 675–691. https://doi.org/10.1016/j.ssci.2007.06.010

Gamil, Y., & Abd Rahman, I. (2023). Studying the relationship between causes and effects of poor communication in construction projects using PLS-SEM approach. *Journal of Facilities Management*, 21(1), 102–148. https://doi.org/10.1108/JFM-04-2021-0039

Gibb, A., & Isack, F. (2003). Re-engineering through pre-assembly: Client expectations and drivers. *Building Research & Information*, *31*(2), 146–160. https://doi.org/10.1080/09613210302000

Gunawardena, T., & Mendis, P. (2022). Prefabricated building systems—Design and construction. *Encyclopedia*, 2(1), 70–95. https://doi.org/10.3390/encyclopedia2010006

Haslam, R.A., Hide, S.A., Gibb, A.G.F., Gyi, D.E., Pavitt, T., Atkinson, S., & Duff, A.R. (2005). Contributing factors in construction accidents. *Applied Ergonomics*, 36(4), 401–415. https://doi.org/10.1016/j.apergo.2004.12.002

Health and Safety Executive (HSE). (2015). *Managing health and safety in construction (CDM Regulations)*. HSE.

Health and Safety Executive (HSE). (2019). *Health and safety at work: Summary statistics*. HSE.

Hughes, P., & Ferrett, E. (2020). *Introduction to health and safety at work* (7th ed.). Routledge.

International Labour Organization (ILO). (2015). *Safety and health in construction*. ILO.

Iqbal, S., Nawaz, M.J., Ali, A., Osman, E., & Hamza, A. (2024). Investigating the impact of project planning on construction project success through the mediating role of risk management and safety climate. *International Journal of Organizational Leadership*, 13(First Special Issue 2024), 119–139. https://doi.org/10.33844/ijol.2024.60426

Jackson, E.N., & Priya, T.S. (2024). Identification and classification of construction-risk factors for ghanaian construction projects: An integrated study with structural equation modelling. *Heliyon*, 10(22), e40397. https://doi.org/10.1016/j.heliyon.2024.e40397

Jang, D.-H., Roh, G.-T., Jeon, C.-H., & Shim, C.-S. (2025). Simulation-based optimization of crane lifting position and capacity using a construction digital twin for prefabricated bridge deck assembly. *Buildings*, 15(3), 475. https://doi.org/10.3390/buildings15030475

Jergeas, G., & Van der Put, J. (2001). Benefits of constructability on construction projects. *Journal of Construction Engineering and Management*, 127(4). https://doi.org/10.1061/(ASCE)0733-9364(2001)127:4(281)

Jkhsi, S. (2020). Evaluation of risk and safety in building structures under construction. *International Journal of Advanced Engineering, Sciences and Applications*, 1(4), 24–28. https://doi.org/10.47346/ijaesa.v1i4.43

Kineber, A.F., Antwi-Afari, M.F., Elghaish, F., Zamil, A.M.A., Alhusban, M., & Qaralleh, T.J.O. (2023). Benefits of implementing occupational health and safety management systems for the sustainable construction industry: A systematic literature review. *Sustainability*, 15(17), 12697. https://doi.org/10.3390/su151712697

Labaran, Y.H., Günal, A.Y., & Saini, G. (2024). Occupational health and safety in the construction industry: A comprehensive review with emphasis on Nigeria. *Turkish Journal of Engineering*, 8(4), 695–711. https://doi.org/10.31127/tuje.1471660

Lingard, H.C., Cooke, T., & Blismas, N. (2012). Designing for construction workers' occupational health and safety: A case study of socio-material complexity. *Construction Management and Economics*, 30(7), 599. https://doi.org/10.1080/01446193.2012.709401

Lingard, H., & Holmes, N. (2001). Understandings of occupational health and safety risk control in small business construction firms: Barriers to implementing technological controls. *Construction Management and Economics*, 19(2), 217–226. https://doi.org/10.1080/01446190010002570

Love, P.E.D., Holt, G.D., Shen, L.Y., Li, H., & Irani, Z. (2002). Using systems dynamics to better understand change and rework in construction project management systems. *International Journal of Project Management*, 20(6), 425–436. https://doi.org/10.1016/S0263-7863(01)00039-4

Ma, J., Yang, Q., Peng, X., & Xia, K. (2025). Review on durability deterioration and mitigation of concrete structures. *Coatings*, 15(9), 982. https://doi.org/10.3390/coatings15090982

Malsane, S., Merin, M.J., Gokul, G.J., Nithin Dev, P., Babu, S.T. (2022). Management of high-volume clashes during clash detection. In: Srivastava, P., Thakur, S.S., Oros, G.I.,

AlJarrah, A.A., & Laohakosol, V. (Eds.) *Mathematical, computational intelligence and engineering approaches for tourism, agriculture and healthcare*. Lecture Notes in Networks and Systems, vol 214. Springer, Singapore. https://doi.org/10.1007/978-981-16-3807-7_23

Mona, M., Mollo, L.G., & Emuze, F. (2025). The connection between workarounds and safety hazards on construction sites. *MATEC Web Conference*, 411, p. 03002. https://doi.org/10.1051/matecconf/202541103002

Muñoz-La Rivera, F., Mora-Serrano, J., & Oñate, E. (2021). Factors influencing safety on construction projects (fSCPs): Types and categories. *International Journal of Environmental Research and Public Health*, 18(20), 10884. https://doi.org/10.3390/ijerph182010884

Nishaant, H., Chellappa, V., Sudhakumar, J.N., & Arunkumar, R. (2025). Identification of challenges influencing project performance in the construction sector: Stakeholder's perspective. *Journal of Legal Affairs and Dispute Resolution in Engineering and Construction,* 17(4). https://doi.org/10.1061/JLADAH.LADR-1378

Olatunji, A.A., & Essan-Ojuri, O. (2018). Factors influencing the buildability of designs in the Nigerian construction industry. *Interim: Interdisciplinary Journal*, 7(2), 16–27.

Osuizugbo, I.C., & Ojelabi, A.R. (2020). Building productionmanagement practice in the construction industry in Nigeria. *Engineering Management in Production and Services*, 12(2), 56–73. https://doi.org/10.2478/emj-2020-0011

Ottenhaus, L.M., Yan, Z., Brandner, R., Leardini, P., Fink, G., & Jockwer, R. (2023). Design for adaptability, disassembly and reuse – A review of reversible timber connection systems. *Construction and Building Materials*, 400, 132823. https://doi.org/10.1016/j.conbuildmat.2023.132823

Rafindadi, A.D., Napiah, M., Othman, I., Mikić, M., Haruna, A., Alarifi, H., & Al-Ashmori, Y.Y. (2022). Analysis of the causes and preventive measures of fatal fall-related accidents in the construction industry. *Ain Shams Engineering Journal*, 13(4), 101712. https://doi.org/10.1016/j.asej.2022.101712

Raheem, A.A., & Hinze, J.W. (2014). Disparity between construction safety standards: A global analysis. *Safety Science*, 70, 276–287. https://doi.org/10.1016/j.ssci.2014.06.012

Rashidi Nasab, A., Malekitabar, H., Elzarka, H., Nekouvaght Tak, A., & Ghorab, K. (2023). Managing safety risks from overlapping construction activities: A BIM approach. *Buildings*, 13(10), 2647. https://doi.org/10.3390/buildings13102647

Reason, J. (1997). *Managing the risks of organizational accidents*. Ashgate.

Sacks, R., Whyte, J., Swissa, D., Raviv, G., Zhou, W., & Shapira, A. (2015). Safety by design: Dialogues between designers and builders using virtual reality. *Construction Management and Economics*, *33*(1), 55–72. https://doi.org/10.1080/01446193.2015.1029504

Sadeghi, H., Cheung, C.M., Yunusa-Kaltungo, A., & Manu, P. (2025). A systematic review of occupational safety and health in modular integrated construction. *Safety Science*, 189, 106897. https://doi.org/10.1016/j.ssci.2025.106897

Saleh, R.A., & Othman, N. (2022). Overview of the causes of accident in construction industry: A comparative perspectives. *International Journal of Academic Research in Economics and Management and Sciences*, 11(4), 1–10. https://dx. doi.org/10.6007/IJAREMS/v11-i4/14253

Sarvari, H., Edwards, D.J., Rillie, I., & Posillico, J.J. (2024). Building a safer future: Analysis of studies on safety I and safety II in the construction industry. *Safety Science*, 178, 106621. https://doi.org/10.1016/j.ssci.2024.106621

Schulte, P.A., Rinehart, R., Okun, A., Geraci, C.L., & Heidel, D.S. (2008). National prevention through design (PtD) initiative. *Journal of Safety Research*, 39(2), 115–121. https://doi.org/10.1016/j.jsr.2008.02.021

Tear, M.J., & Reader, T.W. (2023). Understanding safety culture and safety citizenship through the lens of social identity theory. *Safety Science*, 158, 105993. https://doi.org/10.1016/j.ssci.2022.105993

Terwel, K., & Jansen, S.J.T. (2014). Critical factors for structural safety in the design and construction phase. *Journal of Performance of Constructed Facilities*, 29(3), 04014068. https://doi.org/10.1061/(ASCE)CF.1943-5509.0000560

Toole, T.M. (2005). Increasing engineers' role in construction safety: Opportunities and barriers. *Journal of Professional Issues in Engineering Education and Practice*, 131(3). https://doi.org/10.1061/(ASCE)1052-3928(2005)131:3(199)

Trask, C., & Linderoth, H.C.J. (2023). Digital technologies in construction: A systematic mapping review of evidence for improved occupational health and safety. *Journal of Building Engineering*, 80, 108082. https://doi.org/10.1016/j.jobe.2023.108082

Tu, H., Wei, Z., Bahrami, A., Kahla, N.B., Ahmad, A., & Özkılıç, Y.O. (2023). Recent advancements and future trends in 3D concrete printing using waste materials. *Developments in the Built Environment*, 16, 100187. https://doi.org/10.1016/j.dibe.2023.100187

Vagtholm, R., Matteo, A., Vand, B., & Tupenaite, L. (2023). Evolution and current state of building materials, construction methods, and building regulations in the U.K.: Implications for sustainable building practices. *Buildings*, 13(6), 1480. https://doi.org/10.3390/buildings13061480

Weinstein, M., Gambatese, J.A., & Hecker, S. (2005). Can design improve construction safety? Assessing the impact of a collaborative safety-in-design process. *Journal of Construction Engineering and Management*, 131(10). https://doi.org/10.1061/(ASCE)0733-9364(2005)131:10(1125)

Wong, Q., Sapuan, N.M., & Ali Khan, M.W. (2023). The impact of safety and health towards construction workforce productivity. *Journal of Governance and Integrity*, *6*(1), 504–514. https://doi.org/10.15282/jgi.6.1.2023.9114

Zaryczańska, K., & Karwot, J. (2025). Analysis of occupational safety and health at the construction site based on images acquired from various sources using artificial intelligence. *Safety and Security Advances*, 1(1), 1–17. https://doi.org/10.61093/ssa.1(1).1-17.2025

Zhang, Q., Liu, Z., Yang, S. (2025). Enhancing construction workers' health and safety: Mechanisms for implementing COnstruction 4.0 technologies in construction organizations. *Engineering, Construction and Architectural Management*, 32(13), 68–103. https://doi.org/https://doi.org/10.1108/ECAM-11-2024-1517

Zhang, S., Teizer, J., Lee, J.K., Eastman, C, M., & Venugopal, M. (2013). Building information modeling (BIM) and safety: Automatic safety checking of construction models and schedules. *Automation in Construction*, 29, 183–195. https://doi.org/10.1016/j.autcon.2012.05.006

Zohourian, M., Pamidimukkala, A., Kermanshachi, S., & Almaskati, D. (2025). Modular construction: A comprehensive review. *Buildings*, 15(12), 2020. https://doi.org/10.3390/buildings15122020

7 Stakeholder Roles in Buildability Implementation

7.1 Clients' Roles in Buildability Implementation

In construction, clients are the project owners and play a foundational role in determining whether buildability principles are meaningfully adopted. As project initiators and financiers, clients set the tone for collaboration, innovation, and risk management across the project lifecycle. Within the industry, clients are traditionally categorised into private and public sectors based on ownership structure. Because many clients may be unfamiliar with the technical workings of the construction industry, they often depend on project designers to deliver buildable designs and on professional builders to execute construction works that satisfy project requirements.

By definition, a client is an individual or organisation that, at a particular point in time, has the authority to originate and commission design and construction activities with the objective of enhancing business performance or achieving social goals. In this sense, clients act as sponsors of the entire construction process, providing the most significant perspective on project performance and ensuring that their objectives are met by the project team. This central position places clients among the most influential stakeholders in construction projects. The descriptions of construction clients indicate that their role extends throughout the entire project lifecycle, from project inception to completion. Clients' responsibilities evolve across different project phases, yet their influence consistently has a major impact on project outcomes. Their duties include translating and interpreting project objectives, providing adequate resources, and creating favourable working conditions within the built environment. Consequently, the client's role goes beyond merely defining requirements and arranging project funding; it also involves active commitment to ensuring successful project delivery that satisfies both end-users and other stakeholders.

Project success is closely linked to the client's knowledge, competence, and level of engagement. For timely and effective project delivery, clients must clearly understand their roles, as decisions made at each project stage can either positively or negatively influence buildability, cost, quality, and schedule performance. Clients are uniquely positioned to stimulate innovation, not only by demanding higher construction and process performance or specifying project requirements

DOI: 10.1201/9781003747284-9

but also by establishing and controlling mechanisms that govern communication, coordination, and collaboration among project participants. The likelihood of achieving successful construction outcomes is significantly enhanced when clients assume an active leadership role throughout the project lifecycle. By fostering early stakeholder involvement, supporting collaborative working environments, and encouraging the integration of buildability considerations from the outset, clients can improve project efficiency and reduce risks. Ultimately, effective client leadership and participation contribute to improved project performance and play a vital role in advancing the overall performance of the construction industry.

Improving the buildability of construction projects is a shared responsibility among all key stakeholders, including clients, designers, and builders. This responsibility extends across design development, technical solutions, functional requirements, and process management. However, clients exert a particularly strong influence on buildability outcomes because they initiate, commission, and fund both the design and construction processes, as well as possess the authority to enforce buildability-related decisions. Consequently, the client's understanding of the benefits of improved buildability is critical to its effective implementation. Decisions made by clients during the early phases of the construction process, particularly at the planning and design stages, have far-reaching implications. These early decisions significantly influence construction methods, costs, and risks, yet they are often difficult and costly to modify once construction is underway. As a result, inadequate consideration of buildability at these stages can lead to persistent constructability problems during project execution.

Construction clients/project owners have frequently been identified in the literature as contributors to buildability challenges. For example, several studies report clients' reluctance to invest additional time, funds, and effort during the early stages of projects as a major barrier to improving buildability. Similarly, resistance to adopting formal buildability or constructability programmes has been identified as one of the primary causes of poor buildability performance. Such resistance often limits opportunities for early collaboration, constructability reviews, and design optimisation. These issues may largely stem from a low level of buildability intelligence among construction clients. Buildability intelligence can be defined as the possession of holistic knowledge and awareness of buildability principles, benefits, and implementation strategies. Without this understanding, clients may underestimate the long-term value of early investment in buildability and instead prioritise short-term cost savings, inadvertently increasing construction risks and lifecycle costs.

In recognition of the pivotal role clients play, the following subsections examine in detail the specific roles and responsibilities of clients in the effective implementation of buildability within construction projects.

7.1.1 Strategic Leadership and Vision

Clients exert a significant influence on buildability through the strategic decisions they make at the earliest stages of a construction project, particularly by embedding constructability objectives within project briefs, procurement strategies, and performance metrics. The project brief is a critical document through which clients communicate their expectations, priorities, and constraints to the project team. When buildability requirements, such as ease of construction, safety, efficient sequencing, standardisation, and maintainability, are explicitly stated in the brief, designers and contractors are more likely to integrate these considerations into their proposals and technical solutions. Clear buildability objectives also help align stakeholder efforts and reduce ambiguities that may later lead to design changes or construction inefficiencies.

Procurement strategy is another powerful mechanism through which clients shape buildability outcomes. Traditional procurement approaches that separate design and construction responsibilities often limit the early input of contractors, thereby restricting the incorporation of practical construction knowledge into design development. In such contexts, clients who focus primarily on achieving the lowest initial construction cost tend to encourage competitive pricing at the expense of constructability. This cost-driven mindset may result in complex designs, unrealistic schedules, and inadequate allowances for construction risks, ultimately leading to rework, delays, and increased lifecycle costs.

When clients prioritise lowest initial cost over whole-life value, buildability considerations are frequently compromised. Design decisions may be driven by short-term capital savings rather than long-term performance, leading to solutions that are difficult to construct, maintain, or adapt. Poor buildability often manifests in congested workspaces, unsafe construction methods, inefficient sequencing, and excessive reliance on temporary works. Although such approaches may appear economical at the tender stage, they often result in higher costs during construction and operation due to delays, defects, and increased maintenance requirements.

In contrast, clients who adopt value-based procurement approaches create stronger incentives for buildability. Value-based procurement emphasises whole-life performance, including construction efficiency, safety, durability, operational costs, and sustainability. By assessing bids on a combination of cost, technical quality, buildability, and risk management capability, clients encourage project teams to propose innovative and practical design solutions. Procurement routes such as design–build, early contractor involvement (ECI), and integrated project delivery (IPD) further support this approach by enabling collaboration and knowledge sharing from the early stages of the project.

Performance metrics established by clients also reinforce the importance of buildability. When success criteria extend beyond cost and time to include safety performance, quality outcomes, constructability reviews, and productivity improvements, project teams are motivated to prioritise buildability throughout the project lifecycle. These metrics help shift the focus from transactional project delivery to collaborative value creation.

Ultimately, by embedding constructability objectives into project documentation, adopting value-based procurement strategies, and defining appropriate performance measures, clients can significantly enhance buildability outcomes. Such an approach leads to construction processes that are safer, more efficient, and more predictable, while delivering greater long-term value for both clients and end-users.

7.1.2 Early Stakeholder Engagement

Early stakeholder engagement is a critical enabler of robust buildability implementation, as it allows construction expertise to be integrated into project decision-making from the outset. Buildability challenges often originate during the early design stages, when key decisions regarding layout, structural systems, materials, and construction methods are made. If contractors, suppliers, and construction specialists are engaged only after designs are finalised, opportunities to optimise constructability are limited, and design changes become more expensive and disruptive. Therefore, involving these stakeholders early ensures that practical construction knowledge informs design development before constraints are locked in.

Clients play a central role in facilitating early stakeholder engagement through the procurement strategies they adopt. Traditional design–bid–build procurement typically separates design and construction responsibilities, delaying contractor involvement until after design completion. This fragmentation restricts collaboration and often results in designs that are theoretically sound but difficult to construct. In contrast, alternative procurement routes such as design–build, ECI, and IPD promote collaboration by enabling contractors and key suppliers to contribute during the planning and design stages.

Under design–build arrangements, a single entity assumes responsibility for both design and construction, fostering closer coordination between designers and builders. This integration allows constructability considerations to be addressed continuously, as construction methods, sequencing, and resource availability can be evaluated alongside design development. ECI takes this further by formally engaging contractors during the design phase, even when design and construction contracts remain separate. Contractors can provide input on buildability, cost implications, risk mitigation, and scheduling, enabling more informed and realistic design decisions.

IPD represents the most collaborative form of early stakeholder engagement, as it aligns the interests of clients, designers, contractors, and sometimes key suppliers through shared risks and rewards. IPD encourages transparency, joint decision-making, and collective problem-solving, which are essential for optimising buildability. Through collaborative workshops, constructability reviews, and digital tools such as Building Information Modelling (BIM), stakeholders can identify potential construction issues, resolve clashes, and refine construction sequences before work begins on-site.

The early involvement of suppliers and specialist subcontractors is equally important, particularly for projects that rely on prefabrication, modular construction, or complex building systems. Their expertise helps ensure that designs are compatible with manufacturing processes, logistics constraints, and installation requirements. This alignment reduces material waste, improves productivity, and enhances safety during construction.

Ultimately, early stakeholder engagement enables proactive rather than reactive management of buildability. By allowing construction knowledge to shape design decisions early, clients can minimise costly changes, reduce risks, and improve project outcomes. Such collaborative approaches enhance efficiency, safety, and quality, demonstrating that buildability is best achieved when key stakeholders are engaged from the earliest stages of the project lifecycle.

7.1.3 Resource Allocation and Risk Management

Effective buildability implementation depends not only on technical expertise but also on the strategic allocation of time, financial resources, and risk by clients. Clients occupy a unique position of influence, as their decisions regarding budgeting, scheduling, and contractual arrangements directly shape the behaviours and priorities of project stakeholders. To achieve meaningful buildability outcomes, clients must therefore commit adequate resources to activities and tools that support constructability, while fostering a collaborative contractual environment.

Allocating sufficient time is essential for conducting thorough buildability reviews and constructability workshops. These activities require dedicated periods within the project programme to allow designers, contractors, and specialists to critically evaluate proposed designs, construction methods, and sequencing strategies. When project schedules are compressed to meet unrealistic deadlines, buildability considerations are often rushed or overlooked, leading to unresolved issues that surface during construction. By allowing adequate time for iterative review and coordination, clients enable early identification and resolution of constructability challenges, thereby reducing rework, delays, and disputes during later stages.

Financial investment is equally important in supporting buildability-focused practices. Constructability workshops, specialist consultations, and design optimisation exercises entail upfront costs that some clients may perceive as non-essential. However, these investments typically yield substantial long-term benefits through improved productivity, reduced waste, and enhanced safety performance. Clients must recognise that spending modest resources early in the project can prevent significantly higher costs during construction and operation.

Digital tools, particularly BIM, play a critical role in enhancing buildability. BIM enables stakeholders to visualise construction processes, simulate sequencing, detect clashes, and assess logistical constraints before physical construction begins. However, effective BIM implementation requires investment in software, training, data management, and coordination processes. Clients who mandate and fund appropriate BIM use create a shared digital environment that supports collaboration and informed decision-making, thereby improving constructability and overall project performance.

Beyond time and financial resources, the way clients allocate risk and structure contracts has a profound impact on buildability outcomes. Adversarial contracts that transfer excessive risk to designers or contractors often discourage innovation and proactive problem-solving. In such environments, stakeholders may focus on protecting themselves from liability by adopting conservative or minimal-compliance solutions, rather than proposing buildability-enhancing alternatives. This defensive behaviour undermines collaboration and limits the exploration of more efficient construction approaches.

Table 7.1 Roles of clients in buildability improvement

S/No.	*Roles*
1	Clearly state project objectives
2	Clearly state project priorities
3	Willingness to invest additional money and effort in the early phase of project
4	Find the most cost-effective way to achieve project objectives
5	Use the project objectives as decisive factors in decision-making
6	Consider new design approaches
7	Early involvement of constructors in the design process consideration
8	Consider open construction input as a significant factor in selecting project designers
9	Ensure key personnel allocated to the design team have suitable experience
10	Implement a buildability assessment programme to improve project buildability
11	Commitment to buildability assessment
12	Clients should provide clear requirements on buildability, apart from time/cost/quality targets
13	Lay emphasis on early cost control to workforce
14	Use buildability assessment to achieve project objectives
15	Insists on professionals' participation during major decisions
16	Use project objectives in resolving conflict
17	Retain lessons-learned records from completed projects

Source: Osuizugbo et al. (2023).

Conversely, clients who adopt fair and balanced risk allocation encourage openness, trust, and shared responsibility. Collaborative contractual arrangements, such as partnering agreements or IPD models, align stakeholder incentives and reward collective performance. When risks and rewards are shared equitably, project participants are more willing to suggest innovative construction methods, alternative materials, or sequencing strategies that improve buildability, even if they deviate from traditional practices.

In summary, clients play a crucial role in enabling buildability by allocating sufficient time and financial resources for constructability-focused activities and digital tools, while fostering a non-adversarial contractual environment. These actions create conditions in which stakeholders are motivated to collaborate, innovate, and deliver designs and construction solutions that are safer, more efficient, and more constructible. Roles of clients in improving the buildability of construction projects are summarised and listed in Table 7.1. An understanding of these roles is vital for reducing buildability problems as well as for improving and embedding buildability as a practice in the construction industry.

7.2 Designers' Roles in Buildability Implementation

The construction industry is a vital contributor to economic growth and national development; however, it continues to experience persistent buildability challenges that undermine project performance. Designers translate client objectives into physical solutions and therefore occupy a central position in buildability implementation. Their design decisions directly influence construction complexity, productivity, safety, and long-term maintainability. When buildability considerations are inadequately addressed during the design stage, inefficiencies and risks are transferred downstream into construction, where corrective actions become costly and disruptive.

Several factors contribute to buildability problems within the construction industry. These include the fragmentation of design and construction processes, lack of design standardisation, limited awareness and application of buildability principles, late involvement of construction expertise in design development, and weak integration between designers and builders. Additional challenges such as designers' limited construction experience, insufficient detailing, defective or unclear working drawings, and resistance from clients to buildability practices further exacerbate these issues. Collectively, these shortcomings result in poor workmanship, excessive rework, cost overruns, project delays, disputes among project participants, and, in extreme cases, project abandonment.

Construction projects are executed by a workforce comprising both skilled and unskilled personnel, which intensifies the need for clear, practical, and constructible designs. Poorly considered design solutions place undue pressure on-site personnel, increasing the likelihood of errors, safety incidents, and inefficiencies. As other sectors of the economy rely heavily on the construction industry's outputs, the inability to deliver value through improved buildability has far-reaching economic and social consequences. Enhancing buildability is therefore essential for the construction sector to meet broader development objectives.

Project designers are frequently identified in the literature as significant contributors to buildability problems, high construction costs, and the failure of clients to achieve maximum value from their investments. In particular, designers' limited understanding of buildability concepts and lack of practical construction experience have been cited as major causes of constructability challenges. Questions have also been raised regarding architects' capacity to effectively supervise and manage construction activities, while concerns have been expressed about the quality of services provided by engineering consultants in some contexts. These challenges are often attributed to a low level of buildability intelligence among design professionals, where buildability intelligence refers to the holistic knowledge and practical understanding of constructability principles and their application.

The responsibility for integrating buildability into project design lies primarily with the design team. However, the dominance of traditional procurement methods in the construction industry has reinforced professional silos and adversarial relationships, creating barriers to collaboration. This separation between design and construction has been widely identified as a key reason for the limited application of buildability assessments during design development. As a result, constructability reviews are often overlooked or treated as secondary considerations, despite their potential to prevent downstream construction problems.

Given the sequential nature of design and construction in traditional project delivery systems, it is unsurprising that contractors frequently complain about designs that are impractical, overly complex, or insensitive to site conditions. Such inconsiderate designs increase construction difficulty, reduce productivity, and heighten safety risks. To address these challenges, designers must actively engage with construction knowledge, embrace collaborative working practices, and embed buildability principles throughout the design process. Doing so is essential to improving project outcomes and enhancing the overall performance of the construction industry.

Project designers such as architects, structural engineers, mechanical and electrical engineers, and service engineers, among others, have roles to play in project buildability improvement. These responsibilities cover design, functionality, and process management of construction projects. An understanding of these roles would minimise buildability problems, facilitate buildability improvement, and embed buildability as a practice in construction management. In recognition of the pivotal role designers play, the following subsections examine in detail the specific roles and responsibilities of designers in the effective implementation of buildability within construction projects.

7.2.1 Integration of Construction Knowledge

Designers play a critical role in achieving effective buildability by ensuring that design development reflects practical construction realities. Beyond producing aesthetically pleasing and technically sound designs, designers are responsible for incorporating construction methods, sequencing, access requirements, and temporary works considerations into their design decisions. These aspects determine how a project will be executed on-site and directly influence productivity, safety, cost, and overall project success. When such considerations are embedded early in the design process, construction activities can proceed more efficiently and with fewer disruptions.

Incorporating construction methods into design requires designers to understand how various elements of a project will be built, assembled, and installed. This includes awareness of available construction technologies, material handling processes, equipment requirements, and labour capabilities. For example, structural layouts should accommodate feasible erection methods, and component sizes should reflect transportation and lifting constraints. Designers who account for these factors reduce the likelihood of impractical solutions that require redesign once construction begins.

Construction sequencing is another critical consideration that designers must integrate into design development. The order in which building components are constructed affects site logistics, safety, and scheduling. Designers need to consider how different trades will interact, how workspaces will be shared, and how construction activities can progress without interference. Failure to account for sequencing often leads to congestion on-site, inefficient workflows, and increased safety risks. By aligning design solutions with logical construction sequences, designers can significantly improve site productivity and reduce delays.

Access requirements and temporary works are equally important yet frequently overlooked aspects of buildability. Designers must ensure that adequate access is provided for workers, materials, and equipment throughout construction. This includes considering site layout, storage areas, crane locations, and safe movement paths. Temporary works, such as scaffolding, formwork, shoring, and temporary supports, are essential for safe and effective construction but are often treated as afterthoughts. When designers fail to consider these requirements, contractors may be forced to develop ad hoc solutions, increasing risk, cost, and time pressures.

Effectively addressing these design responsibilities requires more than technical competence; it demands openness to collaboration with contractors, manufacturers, and specialist subcontractors. These stakeholders possess valuable practical knowledge that can inform design decisions and highlight potential constructability issues early. Collaborative practices such as constructability reviews, design coordination meetings, and the use of BIM platforms facilitate knowledge sharing and enable designers to test and refine their solutions.

Designs that ignore practical construction constraints frequently result in rework, delays, and cost overruns. Changes made during construction are typically more expensive and disruptive than those made during design, as they may involve dismantling completed work, renegotiating contracts, and extending project timelines. Moreover, poorly constructed designs can compromise safety and quality, undermining client value and stakeholder confidence.

In summary, designers are central to buildability implementation by integrating construction methods, sequencing, access, and temporary works into design development. Through technical expertise and collaborative engagement, designers can minimise construction risks, reduce waste, and deliver projects that are safer, more efficient, and more cost-effective.

7.2.2 Design Optimisation and Detailing

Robust buildability implementation requires designers to give deliberate and systematic attention to standardisation, modularisation, and repetition throughout the design process. These principles are fundamental to reducing construction complexity, improving productivity, and enhancing overall project performance. When design solutions are overly customised or excessively complex, they place additional demands on construction resources, increase the likelihood of errors, and often lead to higher costs and longer delivery times. By contrast, standardised and modular designs enable more predictable and efficient construction processes.

Standardisation involves the use of consistent dimensions, details, components, and construction methods across a project. Designers who adopt standardised solutions reduce the learning curve for construction teams, as workers become familiar with repeated tasks and details. This familiarity improves productivity, minimises mistakes, and enhances safety. Standardisation also facilitates easier quality control and inspection, as repeated elements can be reviewed and corrected systematically. From a supply chain perspective, standardised components are more readily available, reducing procurement risks and delays associated with custom-made items.

Modularisation further enhances buildability by enabling building elements or systems to be fabricated off-site under controlled conditions and then assembled on-site. Designers play a crucial role in ensuring that modules are appropriately sized, detailed, and coordinated with site constraints such as transportation, lifting capacity, and installation sequences. Modular construction can significantly reduce on-site labour requirements, shorten construction durations, and improve workmanship quality. However, successful modularisation depends on early design decisions and close coordination between designers, manufacturers, and contractors.

Repetition is closely linked to standardisation and modularisation and contributes significantly to constructability. Repetitive design elements allow construction teams to refine processes, optimise resource use, and achieve higher levels of efficiency as work progresses. Repetition also supports safer working environments, as predictable tasks reduce uncertainty and exposure to risk. Designers who intentionally incorporate repetitive layouts and details help create construction processes that are easier to manage and control.

Simplifying details is another critical way designers enhance buildability. Complex junctions, intricate detailing, and unnecessary variations often create confusion on-site and increase the likelihood of misinterpretation. By minimising bespoke components and avoiding unnecessary customisation, designers reduce the need for specialised skills, tools, and fabrication processes. Selecting materials and building systems that are readily available and familiar to the construction industry further supports constructability, as contractors are more likely to have experience with such products and established installation practices.

Clear and coordinated drawings and specifications are essential for translating design intent into effective construction. Poorly coordinated documents, conflicting information, and ambiguous details are major sources of construction errors, rework, and disputes. Designers must ensure that drawings across disciplines are fully aligned and that specifications clearly define materials, tolerances, and performance requirements. The use of digital coordination tools, such as BIM, can significantly improve document clarity and consistency.

In summary, careful attention to standardisation, modularisation, repetition, and design clarity enables designers to deliver solutions that are practical, efficient, and reliable. By simplifying details, reducing bespoke elements, and producing coordinated documentation, designers play a pivotal role in minimising construction errors, controlling costs, and improving overall buildability.

7.2.3 Use of Digital Tools and Feedback Loops

Designers are increasingly leveraging BIM and other digital platforms as essential tools for enhancing buildability and overall project performance. These technologies enable designers to move beyond traditional two-dimensional representations and develop integrated digital models that capture the geometric, functional, and technical characteristics of a project. Through BIM-based workflows, designers can simulate construction processes, visualise sequencing, and evaluate constructability issues long before physical construction begins, thereby reducing uncertainty and risk.

One of the key benefits of BIM is its ability to support clash detection across multiple disciplines, including architecture, structure, and building services. By identifying spatial conflicts between components such as beams, ducts, pipes, and electrical systems during the design stage, designers can resolve issues proactively rather than reacting to problems on-site. Early clash detection significantly reduces rework, delays, and cost overruns, while also improving safety by minimising the need for last-minute design changes during construction.

BIM also enables designers to assess construction sequencing and site logistics through four-dimensional (4D) modelling, which integrates time with the three-dimensional model. This capability allows designers and contractors to simulate the order of construction activities, evaluate access requirements, and identify potential congestion or safety risks. Similarly, five-dimensional (5D) BIM integrates cost data, enabling designers to understand the financial implications of design decisions and assess the cost-effectiveness of alternative solutions. These simulations support informed decision-making and promote designs that are both efficient and constructible.

Beyond BIM, other digital platforms such as digital twins, parametric design tools, and collaborative cloud-based environments further enhance designers' ability to evaluate buildability. These tools facilitate real-time collaboration among project stakeholders, improve coordination, and allow design options to be tested rapidly against construction constraints. As a result, designers can refine solutions iteratively and align design intent more closely with construction realities.

However, effective use of digital tools alone is insufficient without learning from post-construction feedback. Designers must actively engage in post-project evaluations to understand how their designs performed during construction and operation. Feedback from contractors, facility managers, and end-users provides valuable insights into what worked well and what caused difficulties. Issues such as difficult detailing, inefficient sequencing, maintenance challenges, or unexpected site constraints can inform improvements in future projects.

Incorporating lessons learned into organisational knowledge systems helps designers continuously enhance buildability intelligence. By documenting and sharing post-construction experiences, design teams can avoid repeating past mistakes and progressively improve design quality. This feedback loop supports a culture of continuous improvement and strengthens the link between design and construction.

In summary, BIM and digital platforms empower designers to proactively assess constructability, reduce risks, and optimise construction outcomes. When combined with systematic learning from post-construction feedback, these tools enable designers to continuously refine their practices and deliver designs that are more efficient, safer, and better aligned with construction realities. The project designers' roles are very important for improving the buildability of construction projects. A summary of the project designers' roles is shown in Table 7.2.

Table 7.2 Project designers' roles in buildability improvement

S/No	*Project Designers' Role*
1	Enthusiastically seek out for construction input throughout the phases of design process
2	Timely schedule for construction input
3	Communicate fully the design objective to construction
4	Seek for and accept active involvement of constructors throughout the design process
5	Respond fully to every construction input
6	Always seek feedback and enhance the design on the basis of construction outputs
7	Discussing with builders so as to provide design solution
8	Allowing design needs to be coordinated and visualised easily by site workers
9	Designing to permit innovative building methods to be planned by the contractor
10	Studies alternatives and innovative design approaches
11	Often repeat design strategy with construction technique
12	Design with simplicity to minimise difficulty as well as construction scope to the barest minimum
13	Identify and justify every design requirement in surfeit of standard or code practices
14	Design for a system of construction that conserves the environment
15	Design for construction safety (e.g. below ground)
16	Design for interchangeability and sub-assemblies
17	Design to aid the visualisation of completed work
18	Design for locally available and accessible construction materials
19	Specify suitable and robust materials or components
20	Design for sustainable materials in the project
21	Design for the best use of equipment and plant
22	Package design so as to present essential information for efficient contracting, procurement, and construction
23	Define the contractor interface clearly in the design document
24	Allow maximum construction flexibility
25	Co-ordinating and rationalising of design information
26	Use blow-up details to check conflicts in the design (e.g. reinforcements conflicting with building services)

Source: Osuizugbo et al. (2025).

7.3 Roles of Regulatory Bodies in Buildability Implementation

Regulatory bodies play a pivotal role in creating an enabling environment for the effective implementation of buildability in construction projects. Their influence extends well beyond traditional compliance and enforcement functions to encompass policy formulation, technical guidance, industry capacity development, and performance oversight. Through the development of building codes, standards, and guidelines, regulatory authorities establish minimum requirements that promote safe, efficient, and constructible designs, thereby shaping industry practices at a systemic level.

In addition to setting regulatory frameworks, regulatory bodies support buildability through transparent and efficient approval processes. Early engagement with project teams during planning and design stages allows potential constructability challenges to be identified and addressed before construction begins, reducing delays, redesign, and project risk. Regulatory authorities also play a facilitative role by encouraging the adoption of digital tools, such as BIM, which enhance coordination and improve constructability assessment.

Furthermore, regulatory bodies contribute to buildability implementation by promoting innovation, supporting modern construction methods, and investing in industry training and knowledge dissemination. Through monitoring, benchmarking, and feedback mechanisms, they enable continuous improvement across the construction sector. Collectively, these interconnected roles allow regulatory authorities to influence how buildability principles are systematically embedded across the planning, design, approval, and construction stages of projects, ultimately improving project efficiency, safety, and quality. In recognition of the pivotal role regulatory bodies play, the following subsections examine in detail the specific roles and responsibilities of regulatory bodies in the effective implementation of buildability within construction projects.

7.3.1 Policy, Codes, and Standard Development

Regulatory bodies play a fundamental role in shaping buildability outcomes through the formulation and enforcement of building codes, standards, and technical guidelines. These regulatory instruments establish the baseline requirements that govern how buildings are designed and constructed, thereby exerting a direct influence on constructability. When codes and standards explicitly incorporate constructability, safe-by-design principles, and construction efficiency requirements, they encourage designers and project teams to consider practical construction constraints from the earliest stages of project development. Such regulations help ensure that designs are not only compliant but also feasible, safe, and efficient to construct.

By embedding buildability-related provisions into regulatory frameworks, authorities can drive consistency and good practice across the construction industry. For example, requirements relating to access for construction and maintenance, tolerances, material handling, structural simplicity, and coordination of building services can significantly reduce construction complexity and on-site risks. In some jurisdictions, regulatory bodies go further by integrating explicit buildability or constructability assessment criteria into planning and design approval processes. Under such systems, project teams may be required to demonstrate how proposed designs address ease of construction, safety considerations, and productivity improvements. This institutionalisation of buildability ensures that constructability is treated as a core performance criterion rather than an optional or discretionary consideration.

The adoption of performance-based regulations represents an important shift in regulatory philosophy that further enhances buildability implementation. Unlike prescriptive regulations, which dictate specific materials, dimensions, or construction methods, performance-based frameworks focus on achieving defined outcomes related to safety, durability, functionality, and efficiency. This approach provides designers and contractors with greater flexibility to select construction methods and technologies that best suit project conditions while still meeting regulatory objectives. As a result, project teams are better able to optimise designs for constructability without being constrained by rigid regulatory requirements.

Performance-based regulations also create an enabling environment for innovation and the adoption of modern construction technologies. Techniques such as prefabrication, modular construction, and advanced digital design tools often require flexibility in design and approval processes that prescriptive codes may not readily allow. By emphasising performance outcomes rather than specific solutions, regulatory bodies can accommodate these technologies while maintaining appropriate safety and quality benchmarks. This regulatory flexibility encourages continuous improvement and supports productivity gains across the construction sector.

Overall, through the strategic development of codes, standards, and guidelines, combined with a shift toward performance-based regulation, regulatory bodies play a decisive role in embedding buildability principles into construction practice. Their actions promote safer, more efficient, and more innovative construction outcomes, while ensuring consistent quality and industry-wide adoption of best practices.

7.3.2 Approval Processes, Compliance, and Early Engagement

Efficient, transparent, and predictable approval processes play a critical role in supporting buildability by reducing uncertainty and enabling informed decision-making throughout the project lifecycle. Regulatory approval is a key interface between project teams and governing authorities, and the manner in which this process is managed can either facilitate or hinder constructability. When approval procedures are clearly defined, timely, and consistently applied, project teams can plan construction activities with confidence, minimising disruptions caused by regulatory uncertainty.

Regulatory bodies that engage proactively with project teams through pre-application consultations, early design reviews, and advisory services significantly enhance buildability outcomes. Early engagement allows regulators to provide guidance on compliance requirements, site constraints, access provisions, safety considerations, and permissible construction methods before designs are finalised. This collaborative approach enables potential constructability challenges to be identified and addressed during the design stage, when changes are easier and less costly to implement. As a result, project teams can avoid late-stage design revisions that often disrupt construction schedules and inflate project costs.

Pre-application consultations also help align regulatory expectations with project objectives, reducing the likelihood of misunderstandings and conflicting interpretations of requirements. Early design reviews provide an opportunity to assess whether proposed design solutions adequately address construction sequencing, temporary works, and site logistics, which are often overlooked in conventional approval processes. By offering advisory services rather than purely enforcement-focused interactions, regulatory bodies foster a more constructive relationship with industry stakeholders and promote shared responsibility for buildability outcomes.

In contrast, rigid approval systems that rely heavily on late-stage compliance checks can undermine buildability. When regulatory feedback is provided only after detailed designs have been completed, project teams may be forced to implement reactive changes that compromise construction efficiency. Such late interventions often result in redesign, rework, and delays, increasing both direct and indirect project costs. They may also lead to adversarial relationships between regulators and project participants, further hindering collaboration.

Consistent enforcement of regulations is equally important in ensuring that buildability principles are effectively translated from design to construction. Once designs are approved, regulatory bodies must ensure that construction activities align with the approved, constructible solutions. Regular inspections, audits, and compliance monitoring help prevent unauthorised design changes, poor workmanship, and unsafe construction practices. This oversight reinforces accountability and helps maintain the integrity of buildability-focused design decisions.

Furthermore, consistent enforcement reduces quality defects and safety risks by ensuring adherence to established standards and approved methodologies. It also creates a level playing field within the industry, discouraging non-compliance and promoting best practices. When project teams are confident that regulations will be applied fairly and consistently, they are more likely to invest in constructible and compliant solutions from the outset.

In summary, efficient approval processes combined with proactive regulatory engagement and consistent enforcement form a critical foundation for effective buildability implementation. By enabling early identification of constructability issues and ensuring faithful execution of approved designs, regulatory bodies play a vital role in enhancing construction efficiency, safety, and quality.

7.3.3 Digitalisation and BIM Enablement

Regulatory authorities are increasingly promoting buildability through the adoption of digital submission platforms and the integration of BIM requirements into regulatory approval processes. As construction projects become more complex, traditional paper-based or two-dimensional documentation methods are often inadequate for assessing constructability. Digital platforms enable regulators to review coordinated, data-rich models that provide a more comprehensive understanding of design intent, construction methods, and spatial relationships among building components.

BIM-enabled approval processes significantly enhance coordination across architectural, structural, and building services disciplines. By reviewing integrated digital models rather than isolated drawings, regulatory authorities can more effectively identify inconsistencies, clashes, and omissions that may affect constructability. This improved visibility allows regulators to assess whether proposed designs accommodate construction sequencing, access requirements, safety provisions, and maintenance considerations. As a result, potential constructability issues can be addressed during the approval stage rather than emerging during construction.

The use of BIM also improves design clarity and communication between project teams and regulators. Three-dimensional visualisation helps clarify complex design elements that may be difficult to interpret from conventional drawings, reducing misinterpretation and ambiguity. When regulatory feedback is based on a shared digital model, communication becomes more precise and actionable, supporting faster and more accurate design revisions. This clarity contributes to more reliable approvals and smoother transitions from design to construction.

Standardised digital workflows further strengthen buildability implementation by reducing documentation errors and improving consistency. Regulatory authorities that define clear digital submission standards, such as model naming conventions, data structures, and levels of information requirements, help ensure that submissions are complete, coordinated, and comparable. Standardisation reduces the risk of missing or conflicting information, which is a common source of construction errors and rework. It also streamlines the approval process by enabling regulators to review submissions more efficiently and systematically.

Digital platforms also facilitate collaboration among stakeholders by providing shared access to up-to-date project information. Cloud-based environments allow designers, contractors, and regulators to review, comment on, and coordinate design information in real time. This collaborative approach supports early issue resolution and reinforces shared responsibility for buildability outcomes. In addition, digital records created during the approval process provide a valuable reference throughout construction and operation, supporting compliance monitoring and asset management.

Overall, the adoption of digital submission platforms and BIM requirements represents a significant advancement in regulatory support for buildability. By enhancing coordination, improving clarity, and enabling more effective constructability assessment, digitalisation helps regulatory authorities create a more efficient, transparent, and collaborative approval environment. This, in turn, contributes to safer, more efficient, and more buildable construction projects.

7.3.4 Capacity Building, Training, and Professional Development

Beyond their formal regulatory and enforcement functions, authorities play a crucial role in strengthening industry capacity and fostering long-term improvements in buildability across the construction sector. By investing in education, skills development, and knowledge dissemination, regulatory bodies help create an industry environment in which buildability principles are better understood, consistently applied, and continuously improved. These capacity-building initiatives are essential for bridging the gap between regulatory requirements and practical implementation on construction projects.

Training programmes organised or supported by regulatory authorities provide a structured platform for enhancing practitioners' understanding of buildability concepts. Such programmes may target designers, contractors, project managers, inspectors, and other industry professionals, focusing on topics such as constructability principles, safe-by-design practices, modular construction, and digital tools.

By promoting shared learning and common standards of practice, training initiatives help ensure that buildability considerations are embedded across different professional roles and project stages.

Certification schemes and professional accreditation further reinforce buildability implementation by establishing minimum competency standards. Regulatory bodies can collaborate with professional institutions to incorporate buildability-related competencies into licensing and certification requirements. This ensures that practitioners possess not only technical knowledge but also a practical understanding of construction processes, sequencing, and site constraints. Certification schemes also promote accountability and encourage continuous skill enhancement throughout professionals' careers.

Research funding is another important mechanism through which authorities support buildability advancement. By sponsoring research into construction methods, materials, digital technologies, and process optimisation, regulatory bodies contribute to the development of evidence-based solutions to constructability challenges. Research outputs, including guidelines, toolkits, and case studies, provide valuable resources for industry practitioners and inform future regulatory updates.

Continuous professional development (CPD) initiatives further strengthen buildability intelligence by encouraging lifelong learning. As construction technologies and practices evolve, CPD programmes help professionals stay current with emerging trends such as BIM, prefabrication, and sustainable construction methods. Regulatory authorities that mandate or incentivise CPD ensure that industry competencies remain aligned with innovation and best practice.

Collectively, these capacity-building efforts improve buildability intelligence across the construction sector by enhancing knowledge, skills, and professional culture. By aligning competencies with evolving construction technologies and practices, regulatory bodies help create a more capable, innovative, and resilient industry. Ultimately, strengthening industry capacity supports more effective buildability implementation, leading to safer, more efficient, and higher-quality construction outcomes.

7.3.5 Promotion of Innovation and Modern Methods of Construction

Regulators play an increasingly important role in supporting buildability by enabling innovation within the construction industry. Traditional regulatory frameworks are often designed around conventional construction methods and may unintentionally constrain the adoption of new technologies and processes that could improve constructability. To address this challenge, regulatory authorities are adopting more flexible and adaptive approaches, such as pilot projects, regulatory sandboxes, and fast-track approval mechanisms, to encourage innovation while maintaining essential safety and quality standards.

Pilot projects provide a controlled environment in which new construction methods, materials, or systems can be tested and evaluated under real project conditions. By supporting or approving pilot initiatives, regulators allow project teams to demonstrate the feasibility, safety, and performance of innovative solutions, such as prefabricated components or modular building systems. The lessons learned from these pilots can inform future regulatory updates and guide broader industry adoption of proven buildability-enhancing practices.

Regulatory sandboxes further support innovation by allowing temporary exemptions or modified compliance requirements for specific projects that test emerging technologies. Within these frameworks, designers and contractors can explore alternative construction approaches without being constrained by regulations that may not yet account for new methods. Sandboxes encourage experimentation and learning, while still operating under regulatory oversight to ensure that public safety and quality are not compromised.

Fast-track approval processes are particularly beneficial for projects that incorporate prefabrication, modular construction, and other modern methods of construction (MMC). These approaches often require early decisions, specialised manufacturing processes, and precise coordination, all of which benefit from expedited approvals. By streamlining approval procedures for MMC-based projects, regulators reduce uncertainty and delays, making such methods more commercially attractive and viable.

Adapting regulations to accommodate new technologies is critical to reducing barriers to innovation. For example, traditional codes may not fully address off-site manufacturing processes, digital fabrication techniques, or novel materials. Regulators that update standards and guidance in response to technological advances help ensure that regulations remain relevant and supportive of buildability. This adaptability enables project teams to leverage efficiencies offered by MMC, including improved quality control, reduced on-site labour, shorter construction durations, and enhanced safety performance.

Overall, by enabling innovation through pilot projects, regulatory sandboxes, fast-track approvals, and adaptive regulatory frameworks, authorities create a supportive environment for modern construction practices. These initiatives reduce institutional barriers to innovation and encourage the adoption of safer, more efficient, and more buildable construction approaches. In doing so, regulators contribute to improved productivity, sustainability, and resilience within the construction industry.

7.3.6 Integration of Safety, Sustainability, and Buildability

Regulatory bodies play a vital role in ensuring that buildability objectives are aligned with OHS and sustainability requirements, thereby promoting a holistic approach to construction project delivery. Buildability, safety, and environmental performance are closely interconnected, and failure to coordinate these objectives can lead to conflicting requirements that undermine project efficiency and outcomes. Through integrated regulatory frameworks, authorities help ensure that constructability improvements do not compromise worker safety or environmental responsibility.

One of the primary mechanisms through which regulators align buildability and safety is the promotion of safe-by-design principles. Safe-by-design emphasises the identification and mitigation of construction and maintenance hazards during the design stage, rather than relying solely on on-site controls. Regulatory requirements that mandate designers to consider access, lifting operations, temporary works, and maintenance activities encourage design solutions that are both constructible and inherently safer. By embedding safety considerations into design approvals, regulators reduce reliance on reactive safety measures and minimise risks during construction.

Regulatory bodies also ensure that buildability-enhancing construction methods comply with occupational health and safety standards. For example, the adoption of prefabrication and modular construction can significantly reduce on-site hazards by shifting work to controlled factory environments. However, such methods introduce new safety considerations related to transportation, lifting, and assembly. Integrated regulations help address these risks while enabling the benefits of improved buildability and productivity.

In parallel, regulators align buildability with sustainability objectives by embedding environmental performance requirements into construction regulations. Environmentally responsible practices, such as material efficiency, waste reduction, energy performance, and lifecycle assessment, are increasingly integrated into design and construction standards. When sustainability regulations are coordinated with constructability requirements, project teams can adopt solutions that are both practical to build and environmentally sound. For instance, standardised and modular designs not only improve buildability but also reduce material waste and support circular construction practices.

Integrated regulatory approaches prevent conflicts that can arise when safety, sustainability, and buildability are addressed in isolation. Without coordination, sustainability-driven design features may be difficult to construct, or buildability-driven solutions may introduce safety or environmental risks. Regulatory bodies mitigate these conflicts by harmonising requirements across disciplines and encouraging multidisciplinary collaboration during design and approval stages.

In summary, regulatory bodies play a crucial role in aligning buildability with occupational health and safety and sustainability objectives. Through safe-by-design requirements and integrated environmental regulations, authorities ensure that construction projects achieve efficiency and constructability without compromising safety or environmental performance. This holistic regulatory approach supports safer, more sustainable, and more buildable construction outcomes.

7.3.7 Industry Governance, Dispute Reduction, and Risk Management

Clear, consistent, and well-coordinated regulations play a crucial role in supporting buildability by reducing ambiguity, misinterpretation, and disputes that often arise from impractical or poorly defined designs. In the construction industry, regulatory uncertainty can lead to conflicting interpretations of requirements, inconsistent design solutions, and disagreements among project stakeholders. When regulations are fragmented, outdated, or inconsistently enforced, designers and contractors may adopt conservative or defensive approaches that undermine constructability and innovation. By contrast, coherent and well-aligned regulatory frameworks provide clarity and predictability, enabling project teams to develop practical and buildable design solutions with confidence.

Well-defined regulations help ensure that design intent, construction methods, and compliance expectations are clearly understood by all parties. This clarity minimises the risk of late-stage regulatory objections that force design changes during construction, which are a common source of disputes and cost overruns. When regulatory requirements are consistent across different authorities and disciplines, such as planning, building control, safety, and environmental agencies, project teams are better able to coordinate design and construction activities effectively. Such coordination reduces overlaps, contradictions, and gaps that can compromise buildability.

Regulatory bodies also play an important role in dispute reduction by supporting alternative dispute resolution (ADR) mechanisms. Traditional litigation processes are often time-consuming, costly, and adversarial, diverting attention away from project delivery. By promoting ADR approaches such as mediation, adjudication, and expert determination, regulators encourage early and collaborative resolution of conflicts related to compliance, constructability, and design interpretation. These mechanisms help preserve working relationships among stakeholders and support timely project delivery.

In addition to dispute resolution, regulatory bodies promote collaborative governance frameworks that encourage shared responsibility and proactive problem-solving. Collaborative governance emphasises partnership, transparency, and joint decision-making among clients, designers, contractors, and regulators. Through industry forums, technical committees, and stakeholder consultations, regulators facilitate dialogue and alignment on regulatory expectations and buildability objectives. This collaborative approach helps project teams anticipate regulatory challenges and address them collectively rather than reactively.

Furthermore, collaborative governance frameworks support a shift away from blame-oriented cultures toward collective accountability. When stakeholders perceive regulators as partners in achieving safe and buildable outcomes, rather than solely as enforcers, they are more likely to engage openly, share information, and propose innovative solutions. This openness fosters trust and encourages early identification of constructability issues, reducing the likelihood of disputes during construction.

In summary, by providing clear and coordinated regulations, supporting alternative dispute resolution mechanisms, and promoting collaborative governance frameworks, regulatory bodies play a vital role in reducing disputes and enhancing buildability. These efforts create a more predictable, cooperative, and efficient construction environment that supports practical design solutions and successful project delivery.

7.3.8 Monitoring, Benchmarking, and Knowledge Dissemination

Regulators play an important role in enhancing buildability by monitoring construction performance and systematically collecting data related to constructability, safety, quality defects, and project delays. Construction projects generate significant amounts of performance data, yet without structured oversight and analysis, valuable lessons are often lost. Regulatory bodies are uniquely positioned to gather, standardise, and analyse this information at an industry-wide level, providing insights that individual organisations may not be able to obtain independently.

Through inspections, audits, compliance reports, and digital reporting systems, regulators can identify recurring buildability issues such as impractical design details, coordination failures, unsafe construction methods, and inefficient sequencing. Data on defects, rework, accidents, and delays help reveal patterns that indicate underlying constructability weaknesses. When collected consistently across multiple projects, such data provide a robust evidence base for improving regulations, standards, and industry practices.

The development of benchmarks is a key outcome of effective performance monitoring. By establishing reference indicators for productivity, safety performance, defect rates, and construction duration, regulatory authorities enable industry stakeholders to compare project performance against recognised standards. Benchmarks support objective assessment of buildability outcomes and encourage organisations to adopt best practices to improve their performance. They also help clients and project teams set realistic expectations during planning and procurement.

Publishing case studies is another important mechanism through which regulators promote buildability. Well-documented examples of successful projects demonstrate how constructability challenges were addressed through effective design, collaboration, and innovation. Conversely, lessons drawn from problematic projects highlight common pitfalls and their consequences. These case studies provide practical guidance and serve as learning tools for designers, contractors, and clients.

Dissemination of lessons learned ensures that knowledge gained from past projects informs future decision-making. Regulatory bodies can share findings through technical reports, guidelines, workshops, and digital knowledge platforms. This proactive knowledge-sharing role supports evidence-based decision-making and helps the industry move away from repeating avoidable mistakes.

In summary, by monitoring construction performance, developing benchmarks, and disseminating case-based knowledge, regulators foster a culture of continuous improvement. These activities strengthen buildability intelligence across the construction sector and support safer, more efficient, and higher-quality project delivery.

7.3.9 Stakeholder Coordination and Industry Leadership

Regulatory bodies serve a critical role as neutral conveners within the construction industry, fostering coordination and collaboration among diverse project stakeholders, including clients, designers, contractors, and suppliers. Construction projects involve multiple parties with different priorities, responsibilities, and perspectives. Clients often focus on cost, functionality, and timely delivery; designers emphasise aesthetics, technical compliance, and innovation; contractors prioritise constructability, safety, and efficiency; while suppliers are concerned with material availability, quality, and logistics. Without effective coordination, these varying objectives can conflict, leading to design changes, construction delays, cost overruns, and reduced buildability. Regulatory authorities, acting as impartial facilitators, help align these interests by creating forums for dialogue, guidance, and knowledge exchange.

One of the primary ways regulators facilitate coordination is through the establishment of industry forums and technical committees. These platforms provide opportunities for stakeholders to discuss emerging challenges, clarify regulatory expectations, and share experiences regarding buildability practices. By bringing together representatives from different sectors, regulators foster mutual understanding and encourage collaboration, ensuring that decisions regarding design, construction methods, and project delivery are informed by practical, multidisciplinary perspectives. This collaborative approach reduces the likelihood of adversarial relationships, which often arise when parties operate in isolation or under conflicting assumptions.

Stakeholder consultations, another key mechanism, allow regulators to gather input from clients, designers, contractors, and suppliers before implementing new policies, codes, or standards. By actively engaging industry participants in the regulatory development process, authorities ensure that guidelines and requirements are realistic, practical, and supportive of buildability. Such consultations also provide an avenue for stakeholders to voice concerns about potential constraints, enabling regulators to adapt requirements in ways that minimise negative impacts on constructability while still maintaining safety, quality, and compliance.

Regulators also provide leadership by facilitating knowledge sharing and promoting best practices. Through workshops, seminars, and guidance documents, authorities can disseminate lessons learned, case studies, and technical innovations that enhance buildability. This leadership role helps standardise expectations across the industry, enabling project teams to adopt proven strategies for improving construction efficiency, reducing errors, and enhancing safety.

Furthermore, regulatory bodies' neutrality is essential in mediating conflicts and fostering trust among stakeholders. When regulators act as unbiased conveners, they provide a safe environment for open dialogue and negotiation. Stakeholders are more likely to engage honestly, share information, and work collaboratively toward solutions that improve constructability, rather than adopting defensive or adversarial stances.

In summary, by acting as neutral conveners, regulatory authorities facilitate coordination, knowledge exchange, and collaborative problem-solving across the construction industry. Through forums, technical committees, and stakeholder consultations, they align diverse interests toward shared buildability goals, reduce conflicts, and foster an environment in which practical, safe, and efficient construction practices can thrive. This role is essential for ensuring that regulatory frameworks support, rather than hinder, the implementation of buildability principles across the project lifecycle.

Regulatory bodies play a central role in embedding buildability principles across the construction lifecycle through a combination of integrated functions. By developing policies, codes, and standards, they establish clear expectations for constructible, safe, and efficient designs. Through streamlined approvals and early engagement, they help project teams identify and address constructability challenges before construction begins, reducing delays and rework. Digital enablement, particularly through BIM and standardised submission platforms, enhances coordination, improves design clarity, and allows regulators to assess constructability more effectively. Capacity-building initiatives, including training, certification, and professional development, strengthen industry competencies, while support for innovation, through pilot projects, regulatory sandboxes, and MMC, encourages safer and more efficient approaches. Finally, monitoring performance and disseminating lessons learned fosters continuous improvement. Collectively, these leadership and governance functions ensure higher efficiency, improved safety, better quality, and overall enhanced construction industry performance.

7.4 Chapter Conclusion

This chapter has underscored that successful buildability implementation is not the responsibility of a single party but a collective endeavour requiring coordinated action among clients, designers, and regulatory bodies. Buildability thrives when strategic intent, technical expertise, and governance frameworks are aligned toward a common objective: delivering projects that are efficient, safe, cost-effective, and sustainable. The roles examined throughout this chapter demonstrate that each stakeholder exerts a distinct yet interconnected influence on project outcomes.

Clients emerge as the primary drivers of buildability through their strategic leadership and vision. By setting clear expectations for constructability, allocating appropriate resources, and promoting early stakeholder engagement, clients establish the conditions necessary for buildability to flourish. Their decisions regarding procurement strategies, risk allocation, and project governance significantly shape collaboration dynamics and the integration of construction knowledge from the outset. When clients actively champion buildability, they create an enabling environment that encourages innovation, proactive risk management, and value-driven design solutions.

Designers, on the other hand, translate strategic intent into practical and constructible solutions. Their role extends beyond aesthetic and functional considerations to include the integration of construction knowledge, optimisation of detailing, and anticipation of site constraints. The effective use of digital tools, particularly BIM and other collaborative platforms, facilitates coordination, reduces errors, and strengthens feedback loops between design and construction teams. By embracing interdisciplinary collaboration and constructability principles, designers act as critical agents in reducing rework, mitigating risks, and enhancing overall project performance.

Regulatory bodies play a foundational governance role in shaping the broader ecosystem within which buildability is implemented. Through the development of policies, codes, and standards, they establish benchmarks that encourage safe, sustainable, and buildable design practices. Streamlined approval processes, digitalisation initiatives, and BIM enablement further enhance efficiency and transparency. Moreover, their commitment to capacity building, professional development, and the promotion of innovation, including MMC, ensures that industry practices evolve in line with technological advancements and societal needs. By integrating safety, sustainability, and buildability into regulatory frameworks, these bodies foster holistic project delivery approaches that balance compliance with performance.

This chapter also highlighted the importance of governance mechanisms such as dispute reduction strategies, risk management frameworks, and performance monitoring systems. Regulatory leadership in benchmarking and knowledge dissemination strengthens industry learning and continuous improvement. Furthermore, effective stakeholder coordination, supported by clear communication channels and shared objectives—minimises fragmentation and promotes collective accountability.

In conclusion, buildability implementation is a systemic process that depends on leadership, collaboration, and institutional support. Clients provide direction and resources, designers operationalise constructability principles, and regulatory bodies create the enabling environment that sustains industry-wide standards and innovation. When these stakeholders act in synergy, buildability transitions from a theoretical concept to a practical, measurable contributor to project success and long-term industry resilience.

Bibliography

Abdulqader, D.N., & Atrushi, D.S. (2022). Evaluation and assessment of existing design codes and standards for building construction: A state of the art. *Aro-The Scientific Journal of Koya University*, 10(2), 106–123. https://doi.org/10.14500/aro.10983

Aina, O.O., & Wahab, A.B. (2011). An assessment of buildability problems in the Nigerian construction industry. *Global Journal of Research Engineering,* 11(2), 42–52.

Aiyetan, O.A., Smallwood, J.J., & Shakantu, W. (2014). Influence of client understanding on quality of design. In: *5th International conference proceedings on engineering, project and production management*, pp. 244–253, November 26–28, Port Elizabeth, South Africa. https://doi.org/10.32738/CEPPM.201411.0025

AlBarami, M.A., Thiruchelvam, S., & Ibrehem, A.S. (2020). Factors affecting the client's performance in the public construction projects: Case study of Oman. *International Journal of Sustainable Construction Engineering and Technology*, 11(2), 248–255. https://doi.org/10.30880/ijscet.2020.11.02.031

Alharthi, A., Soetanto, R., & Edum-Fotwe, F. (2014). Revisiting client roles and capabilities in construction procurement. Available at: https://www.irbnet.de/daten/iconda/CIB_DC27722.pdf (accessed March 12, 2023).

Aniekwu, N.A., Igboanugo, C.A., & Onifade, M.K. (2015). Critical 1ssues in reforming the Nigerian construction industry. *British Journal of Applied Science and Technology*, 5(3), 321–332.

Assaf, M., Hussein, M., Abdelkhalek, S., & Zayed, T. (2023). A multi-criteria decision-making model for selecting the best project delivery systems for offsite construction projects. *Buildings*, 13(2), 571. https://doi.org/10.3390/buildings13020571

Aydın, Z., & Yardımcı, O. (2024). Regulatory sandboxes and pilot projects: Trials, regulations, and insights in energy transition. *Engineering Science and Technology, an International Journal*, 56, 101792. https://doi.org/10.1016/j.jestch.2024.101792.

Barham, D., Matarneh, S.T., Tezel, A., Hasan, S., & Mereb, W. (2025). Opportunities and challenges of digital twin applications in the Middle East construction industry. *Urbanization, Sustainability and Society*, 2(1), 231–258. https://doi.org/10.1108/USS-07-2024-0041

Bello, S.A., Oyedele, L.O., Akinade, O.O., Bilal, M., Delgado, J.M.D., Akanbi, L.A., Ajayi, A.O., & Owolabi, H.A. (2021). Cloud computing in construction industry: Use cases, benefits and challenges. *Automation in Construction*, 122, 103441. https://doi.org/10.1016/j.autcon.2020.103441.

Butt, A., Naaranoja, M., & Savolainen, J. (2016). Project change stakeholder communication. *International Journal of Project Management*, 34(8), 1579–1595. https://doi.org/10.1016/j.ijproman.2016.08.010.

Chigangacha, P.S., & Haupt, T.C. (2017). Effectiveness of client involvement in construction projects: A contractor perspective. In: *Proceedings: 11th Built Environment Conference*, pp. 249–266, August 6–8, Durban. ISBN: 978-0-620-76406-3.

Doan, D.T., Atencio, E., Muñoz La Rivera, F., & Alnajjar, O. (2025). A systematic literature review of building information modelling (BIM) and offsite construction (OSC) integration: Emerging technologies and future trends. *Applied Sciences*, 15(18), 9981. https://doi.org/10.3390/app15189981

Dosumu, O., & Aigbavboa, C. (2018). Conventional versus integrated design and construction process. In: *Sustainable design and construction in Africa* (1st ed.). Routledge. eBook ISBN 9781351212205. https://doi.org/10.1201/9781351212205

Emerson, K., Nabatchi, T., & Balogh, S. (2012). An integrative framework for collaborative governance. *Journal of Public Administration Research and Theory*, 22(1), 1–29. https://doi.org/10.1093/jopart/mur011

Ferreira, M.D.A.V., Vaz Morgado, C.D.R., & Lins, M.P.E. (2024). Organizations and stakeholders' roles and influence on implementing sustainability requirements in construction projects. *Heliyon*, 10(1), e23762. https://doi.org/10.1016/j.heliyon.2023.e23762.

Galaz-Delgado, E.I., Herrera, R.F., Atencio, E., Muñoz-La Rivera, F., & Biotto, C.N. (2021). Problems and challenges in the interactions of design teams of construction projects: A bibliometric study. *Buildings*, 11(10), 461. https://doi.org/10.3390/buildings11100461

Giorgi, S., Atta, N., Dalla Valle, A., Viscuso, S., Lavagna, M., & Talamo, C.M.L. (2025). Pilot projects to put reuse and remanufacturing into practice in the tertiary building sector. *Sustainability*, 17(22), 10374. https://doi.org/10.3390/su172210374

Greenhalgh, B., Squires, G., & Mahamadu, A.M. (2021). Chapter 2: Clients of the construction industry. In *Construction Procurement* (pp. 21–36, Edition: 1). Routledge. https://doi.org/10.1201/9781003155355-2

He, H., Ismail, A.S., & Nita Bte Ali, K. (2026). Systematic literature review on the integration of building information modelling and prevention through design (PtD) for safety risk management. *Journal of Asian Architecture and Building Engineering*, 1–16. https://doi.org/10.1080/13467581.2025.2611515

Hwabamungu, B., & Shepherd, P. (2024). The influence of stakeholder involvement in the adoption of digital technologies in the UK construction industry. *Informatics*, 11(4), 97. https://doi.org/10.3390/informatics11040097

Ibrahim, A., Zayed, T., & Lafhaj, Z. (2024). Enhancing construction performance: A critical review of performance measurement practices at the project level. *Buildings*, 14(7), 1988. https://doi.org/10.3390/buildings14071988

Jarkas, A.M. (2010). Analysis and measurement of buildability factors affecting edge formwork labour productivity. *Journal of Engineering Science and Technology Review*, 3(1), 142–150.

Kiani Mavi, R., Gengatharen, D., Kiani Mavi, N., Hughes, R., Campbell, A., & Yates, R. (2021). Sustainability in construction projects: A systematic literature review. *Sustainability*, 13(4), 1932. https://doi.org/10.3390/su13041932

Kilinc, N., Ozturk, G.B., & Yitmen, I. (2015). The changing role of the client in driving innovation for design-build projects: stakeholders' perspective. *Procedia Economics and Finance*, 21, 279–287.

Lam, P.T.I., & Wong, F.W.H. (2009). Improving building project performance: How buildability benchmarking can help. *Construction Management and Economics*, 27(1), 41–52.

Larsen, G.D., & Whyte, J. (2013). Safe construction through design: perspectives from the site team. *Construction Management and Economics*, 31(6), 675–690. https://doi.org/10.1080/01446193.2013.798424

Lee, S. (2022). When tensions become opportunities: Managing accountability demands in collaborative governance. *Journal of Public Administration Research and Theory*, 32(4), 641–655. https://doi.org/10.1093/jopart/muab051

Lu, Y., Gong, P., Tang, Y., Sun, S., & Li, Q. (2021). BIM-integrated construction safety risk assessment at the design stage of building projects. *Automation in Construction*, 124, 103553. https://doi.org/10.1016/j.autcon.2021.103553.

Memić, N., Tadayon, A., Olsson, N.O.E., Wondimu, P.A., Ishtiaque, T.A., & Lædre, O. (2023). Exploring pre-construction activities in infrastructure projects that can benefit from contractor involvement. *Buildings*, 13(10), 2569. https://doi.org/10.3390/buildings13102569

Mousavi, Y., Gharineiat, Z., Karimi, A.A., McDougall, K., Rossi, A., & Gonizzi Barsanti, S. (2024). Digital twin technology in built environment: A review of applications, capabilities and challenges. *Smart Cities*, 7(5), 2594–2615. https://doi.org/10.3390/smartcities7050101

Mydin, S.H., Zin, R.M., Abdi-Majid, M.Z., Zahidi, M., & Memon, A.H. (2011). Buildability attributes at design phase in Malaysian building construction. *International Journal of Sustainable Construction Engineering & Technology*, 2(1), 24–43.

Nusriadi, L., Avianti, I., Tanzil, N.D., & Parikesit, D. (2023). The collaborative governance elements contributing to implementing public-private partnerships: A systematic literature review. *Journal of Namibian Studies: History Politics Culture*, 33(2), 4473–4489. https://doi.org/10.59670/jns.v33i.1155

Okoh, O.F., Ukpoju, E.A., Otakwu, A., Ayoola, V.B., & Enyejo, L.A. (2024). Construction management: Some issues in the construction project. *Engineering Heritage Journal (GWK)*, 8(1), 42–50. https://doi.org/10.26480/gwk.01.2024.42.50

Orvik, C.P., Albrechtsen, E., Kongsvik, T., Holen, S.M., Bertheussen, L.E., & Moltubakk, S.T. (2026). What is safety performance? A systematic review of conceptualizations in the construction safety research. *Safety Science*, 193, 107025. https://doi.org/10.1016/j.ssci.2025.107025.

Osuizugbo, I.C. (2020a). The need for and benefit of buildability analysis: Nigeria as a case study. *Journal of Engineering, Design and Technology*, 19(5), 1207–1230. https://doi.org/10.1108/JEDT-08-2020-0338

Osuizugbo, I.C. (2020b). Improving the performance of building construction firms through addressing the gap of building production management: A new production model approach. *Journal of Engineering, Project, and Production Management*, 10(1), 50–63. https://doi.org/10.2478/jeppm-2020-0007

Osuizugbo, I.C. (2025). Buildability-led procurement: A new model for future construction projects delivery. In: *Construction Innovation: Information, Process, Management*. https://doi.org/10.1108/CI-10-2023-0272.

Osuizugbo, I.C., Aina, O.O., & Onokwai, A.O. (2023a). Roles of clients in buildability improvement in Nigeria: An exploratory factor analysis. *Built Environment Project and Asset Management*, 13(6), 862–879. https://doi.org/10.1108/BEPAM-01-2023-0007

Osuizugbo, I.C., & Ojelabi, R.A. (2020). Building production management practices in the construction industry in Nigeria. *Engineering Management in Production and Services*, 12(2), 56–73. https://doi.org/10.2478/emj-2020-0011

Osuizugbo, I.C., Okolie, K.C., & Oshodi, O.S. (2022). Factors supporting the implementation of buildability assessment as a tool for buildability improvement. *Journal of Engineering, Design and Technology.* https://doi.org/10.1108/JEDT-12-2021-0738

Osuizugbo, I.C., Okolie, K.C., Oshodi, O.S., & Oyeyipo, O.O. (2023b). Buildability in the construction industry: A systematic review. *Construction Innovation: Information Process Management*, 23(5), 1300–1322. https://doi.org/10.1108/CI-05-2022-0112

Osuizugbo, I.C., Onokwai, A.O., & Faleti, O.O. (2025). Principal factor analysis of project designers' roles in improving construction projects buildability in Nigeria. *Journal of Engineering, Design and Technology*, 23(1), 88–107. https://doi.org/10.1108/JEDT-02-2023-0067

Othman, A.A.E. (2011). Improving building performance through integrating constructability in the design process. *Organization, Technology and Management in Construction: An International Journal*, 3(2), 333–347.

Panas, A., & Pantouvakis, J.P. (2010). An empirical framework for the performance-based evaluation of health and safety's contribution to sustainable construction. *Organization, Technology and Management in Construction: An International Journal*, 2(2), 174–182.

Parracho, D.F.R., Nour El-Din, M., Esmaeili, I., Freitas, S.S., Rodrigues, L., Poças Martins, J., Corvacho, H., Delgado, J.M.P.Q., & Guimarães, A.S. (2025). Modular construction in the digital age: A systematic review on smart and sustainable innovations. *Buildings*, 15(5), 765. https://doi.org/10.3390/buildings15050765

Peter, M. (2024). Understanding building regulations: Ensuring safety and compliance. *Journal of Architectural Engineering Technology*, 13(3), 1000382.

Rajapakshe, W. (2025). Critical risk factors influencing the management of disruptions in construction projects: Insights from recent challenges in Sri Lanka. *Social Sciences & Humanities Open*, 12, 101748. https://doi.org/10.1016/j.ssaho.2025.101748.

Ryd, N. (2014). Construction clients challenges - Emphasizing early stages. *Procedia – Social and Behavioral Sciences*, 119, 134–141.

Saka, A., Taiwo, R., Saka, N., Salami, B.A., Ajayi, S., Akande, K., & Kazemi, H. (2024). GPT models in construction industry: Opportunities, limitations, and a use case validation. *Developments in the Built Environment*, 17, 100300. https://doi.org/10.1016/j.dibe.2023.100300.

Shahzad, M., Shafiq, M.T., Douglas, D., & Kassem, M. (2022). Digital twins in built environments: An investigation of the characteristics, applications, and challenges. *Buildings*, 12(2), 120. https://doi.org/10.3390/buildings12020120

Sobieraj, J., & Metelski, D. (2022). Project risk in the context of construction schedules—combined Monte Carlo simulation and time at risk (TaR) approach: Insights from the Fort Bema housing estate complex. *Applied Sciences*, 12(3), 1044. https://doi.org/10.3390/app12031044

Teo, Y.H., Yap, J.H., An, H., Yu, S.C.M., Zhang, L., Chang, J., & Cheong, K.H. (2022). Enhancing the MEP coordination process with BIM technology and management strategies. *Sensors*, 22(13), 4936. https://doi.org/10.3390/s22134936

Trigunarsyah, B. (2007). Project designers' role in improving constructability of Indonesian construction projects. *Construction Management and Economics*, 25, 207–215. https://doi.org/10.1080/01446190600830649

Tuhaise, V.V., Tah, J.H.M., & Abanda, F.H. (2023). Technologies for digital twin applications in construction. *Automation in Construction*, 152, 104931. https://doi.org/10.1016/j.autcon.2023.104931.

Tumpa, R.J., & Naeni, L. (2025). Improving decision-making and stakeholder engagement at project governance using digital technology for sustainable infrastructure projects. *Smart and Sustainable Built Environment*, 14(4), 1292–1329. https://doi.org/10.1108/SASBE-10-2024-0451.

Voets, J., Brandsen, T., Koliba, C., & Verschuere, B. (2021). Collaborative governance. In: *Oxford research encyclopedia of politics*. Available at: https://oxfordre.com/politics/view/10.1093/acrefore/9780190228637.001.0001/acrefore-9780190228637-e-1419 (accessed Janurary 17, 2026).

Voordijk, H., & olde Scholtenhuis, L. (2024). Technological mediation and 3D visualizations in construction engineering practice. *AI & Society*, 39(3), 207–220. https://doi.org/10.1007/s00146-022-01420-3

Waqar, A., Al Hajj, F., Danish, M., Khan, M.B., & Houda, M. (2025). Evaluating the influence of modern construction methods on sustainable construction success. *Cleaner and Responsible Consumption*, 19, 100327. https://doi.org/10.1016/j.clrc.2025.100327

Welch, S., & Mann, R. (2001). The development of a benchmarking and performance improvement resource. *Benchmarking: An International Journal*, 8(5), 431–452. https://doi.org/10.1108/EUM0000000006387

Wong, F.W.H., De Saram, D.D., Lam, P.T.I., & Chan, D.W.M. (2006a). A compendium of buildability issues from the viewpoints of construction practitioners. *Architectural Science Review*, 49(1), 81–90. https://doi.org/10.3763/asre.2006.4910

Wong, F.W.H., Lam, P.T.I., Chan, A.P.C., & Chan, E.H.W. (2006b). A review of buildability performance in Hong Kong and strategies for improvement. *Surveying and Built Environment*, 17(2), 37–48.

Wong, F.W.H., Lam, P.T.I., Chan, E.H.W., & Shen, L.Y. (2007). A study of measures to improve constructability. *International Journal of Quality & Reliability Management*, 24(6), 586–601. https://doi.org/10.1108/02656710710757781

Wong, F.W.H., Lam, P.T.I., Chan, E.H.W., & Wong, F.K.W. (2011). Factors affecting buildability of building designs. *Canadian Journal of Civil Engineering*, 33(7), 795–806. https://doi.org/10.1139/L06-022.

Part III

Education and Capacity Building

8 Education and Training for Enhanced Buildability

8.1 Key Competencies for Enhancing Buildability in Construction

Enhancing buildability in construction projects depends largely on the competencies of professionals involved across the project lifecycle. Buildability-oriented competence enables the effective translation of design concepts into practical, efficient, and economically viable construction solutions. These competencies are discussed in the following subheadings.

8.1.1 Foundational Competencies

One of the foundational competencies for enhancing buildability in construction is comprehensive technical construction knowledge. This competency encompasses a thorough understanding of construction materials, methods, technologies, and sequencing, alongside an appreciation of site logistics, health and safety requirements, and environmental constraints. Such knowledge enables construction professionals to evaluate how design decisions translate into on-site activities and to foresee practical challenges that may arise during project execution.

A strong grasp of material properties and construction techniques allows practitioners to assess the suitability of design specifications in relation to availability, durability, constructability, and cost. Understanding construction sequencing and workflow is equally critical, as it enables professionals to plan efficient construction processes, avoid trade clashes, and minimise disruptions on-site. Awareness of site logistics, including access, storage, equipment movement, and labour deployment, further supports realistic planning and efficient resource utilisation.

When applied during the early design stages, comprehensive technical construction knowledge empowers professionals to proactively identify potential constructability issues, such as complex detailing, impractical tolerances, or inefficient assembly methods. Early identification allows for timely design modifications or the proposal of alternative construction approaches that simplify execution. These informed interventions can significantly enhance ease of construction, reduce material waste, shorten construction durations, and lower overall project costs.

DOI: 10.1201/9781003747284-11

Moreover, integrating safety considerations into technical decision-making contributes to safer construction practices by eliminating hazards through design rather than relying solely on-site controls. By bridging the gap between design intent and construction reality, comprehensive technical construction knowledge plays a pivotal role in improving buildability, enhancing project performance, and ensuring successful construction outcomes.

8.1.2 Design–Construction Integration

Another critical competency for enhancing buildability in construction is design–construction integration, which emphasises the seamless alignment of design intent with construction practicality. This competency involves the ability of construction professionals to interpret design documents, such as drawings, specifications, and models, through a construction-oriented lens, while also enabling designers to incorporate construction knowledge into design development. Effective design–construction integration ensures that decisions made during the design phase are informed by realistic construction methods, site conditions, and resource constraints.

Construction professionals who possess this competency can provide valuable input on buildable details, material selection, tolerances, and construction sequencing. Their early involvement allows potential constructability issues, such as overly complex geometries or impractical detailing, to be identified and resolved before construction begins. At the same time, designers who understand construction constraints are better positioned to develop solutions that respect project budgets, schedules, and safety requirements without compromising functional or aesthetic objectives.

Strong design–construction integration also relies on collaborative processes and communication mechanisms, including design reviews, constructability assessments, and the use of shared digital platforms such as building information modelling (BIM). These practices facilitate information exchange, improve coordination among disciplines, and promote collective problem-solving. As a result, conflicts between design and construction are minimised, and design intent is more accurately realised on-site.

Ultimately, effective design–construction integration contributes significantly to improved buildability by reducing design-related errors, minimising change orders, and preventing construction delays. By fostering mutual understanding and collaboration between designers and constructors, this competency supports more efficient project delivery, enhanced quality outcomes, and overall project success.

8.1.3 Interdisciplinary Communication and Collaboration

Interdisciplinary communication and collaboration play a vital role in enhancing buildability by ensuring effective coordination among architects, engineers, contractors, and other project stakeholders. Construction projects involve multiple disciplines with distinct perspectives, priorities, and technical languages, and buildability challenges frequently emerge at the interfaces between these disciplines. The ability to communicate clearly and consistently across professional boundaries is therefore essential for identifying and addressing potential constructability issues at an early stage.

Effective interdisciplinary communication enables the timely exchange of information related to design intent, construction methods, sequencing, and site constraints. When stakeholders actively share knowledge and raise concerns during the design and planning phases, conflicts such as incompatible systems, spatial clashes, or impractical detailing can be resolved before they impact construction. This proactive coordination reduces the likelihood of misunderstandings, rework, and costly delays during project execution.

Collaboration also requires the ability to negotiate trade-offs between competing project objectives, such as cost, time, quality, safety, and aesthetics. Through collaborative decision-making, stakeholders can collectively evaluate alternatives and select solutions that best support buildability while meeting overall project goals. Strong teamwork fosters mutual respect and trust, encouraging stakeholders to take shared responsibility for buildability outcomes rather than operating in isolated silos.

Furthermore, interdisciplinary collaboration supports integrated project delivery approaches, such as design–build and integrated project delivery, which emphasise early involvement of key stakeholders and collective problem-solving. These approaches enhance alignment between design and construction processes and promote a culture of cooperation. By strengthening interdisciplinary communication and collaboration, construction teams can significantly improve buildability, leading to more efficient project delivery, higher quality outcomes, and reduced project risk.

8.1.4 Problem-Solving and Critical Thinking Skills

Problem-solving and critical thinking skills are central to enhancing buildability in construction due to the complex, uncertain, and dynamic nature of construction projects. Each project presents unique technical, environmental, and organisational challenges that require professionals to analyse situations systematically, anticipate potential issues, and develop practical solutions. These skills enable practitioners to move beyond routine decision-making and adopt innovative approaches that improve the efficiency and feasibility of construction processes.

Effective problem-solving involves the ability to evaluate multiple design and construction alternatives while considering constraints related to time, cost, quality, safety, and sustainability. Construction professionals with strong critical thinking skills can assess the implications of design choices, construction methods, and sequencing strategies, allowing them to select options that enhance buildability and reduce unnecessary complexity. This capability is particularly important during the early stages of project development, where informed decisions can have a significant impact on overall project outcomes.

Competence in constructability analysis enables professionals to systematically review designs to identify potential construction difficulties, inefficiencies, or risks before construction begins. Similarly, value engineering skills support the evaluation of alternative materials, methods, and systems to achieve required performance at optimal cost without compromising quality or safety. Risk assessment further strengthens buildability by identifying and mitigating potential construction-related risks, such as delays, safety hazards, or supply chain disruptions.

Together, problem-solving and critical thinking foster adaptive and resilient project teams capable of responding effectively to unforeseen challenges during construction. By applying analytical reasoning and evidence-based decision-making, practitioners can optimise construction processes, minimise rework, and enhance overall project performance, ultimately contributing to more buildable and successful construction projects.

8.1.5 Digital and Technological Competence

Digital and technological competence has become increasingly important in enhancing buildability within modern construction practice. As construction projects grow in scale and complexity, digital tools provide critical support for integrating design and construction processes and improving decision-making. Proficiency in BIM, in particular, enables professionals to develop coordinated, data-rich digital representations of projects that facilitate collaboration across disciplines.

BIM supports early detection of design conflicts through clash detection, allowing potential constructability issues between architectural, structural, and building services systems to be identified and resolved during the design phase rather than on-site. This proactive approach significantly reduces rework, delays, and additional costs during construction. Furthermore, BIM-based visualisation enhances understanding of complex design elements and construction details, enabling stakeholders to better assess buildability and propose practical improvements.

Digital collaboration platforms further strengthen coordination by enabling real-time information sharing, version control, and transparent communication among project participants. These platforms reduce information silos and ensure that all stakeholders are working with consistent and up-to-date project data. In addition, construction simulation and 4D modelling tools allow teams to visualise construction sequences and site logistics, supporting more effective planning of workflows, resource allocation, and safety measures.

The ability to evaluate alternative construction methods, materials, and sequences using digital tools also supports evidence-based decision-making. By simulating different scenarios, professionals can assess the buildability implications of design choices and select options that optimise efficiency and risk management. Overall, digital and technological competence enhances buildability by improving coordination, reducing uncertainty, and enabling more informed, collaborative, and efficient construction practices.

8.1.6 Practical Experience and Contextual Awareness

Practical experience and contextual awareness underpin all other competencies required for enhancing buildability in construction. While technical knowledge and analytical skills are essential, their effective application depends largely on an understanding of real-world project environments. Practical experience enables construction professionals to appreciate how theoretical principles translate into on-site realities, including constraints related to time, cost, resources, and site conditions.

Contextual awareness involves a comprehensive understanding of regulatory and statutory requirements, such as building codes, safety regulations, and environmental standards, which directly influence design decisions and construction methods. Awareness of labour availability and skill levels is equally important, as buildability is affected by the capacity of the workforce to execute specific construction techniques efficiently and safely. In addition, knowledge of supply chain limitations, including material availability, lead times, and logistics, allows professionals to select construction solutions that are realistic and achievable within project constraints.

Local construction practices and cultural norms further shape buildability outcomes. Familiarity with regional construction methods, climate conditions, and industry standards enables professionals to tailor designs and construction strategies to local contexts, thereby reducing uncertainty and risk. Such contextual insight supports more informed decision-making and enhances the feasibility of proposed construction solutions.

Continuous professional development and experiential learning are essential for maintaining and advancing buildability competence in a rapidly evolving construction industry. On-the-job training, site exposure, mentoring, and participation in professional education programmes allow practitioners to update their skills and adapt to new technologies and practices. Collectively, practical experience and contextual awareness strengthen the effective application of buildability principles, contributing to improved efficiency, enhanced safety, higher quality outcomes, and overall project success.

8.2 Curriculum Design for Buildability Competence in Construction

In modern construction education, curriculum design is a critical determinant in developing graduates who are capable of enhancing buildability in construction projects. Buildability, broadly defined as the ease and efficiency with which a structure can be constructed while meeting quality, cost, safety, and sustainability goals, requires professionals who possess a combination of technical knowledge, practical skills, and critical decision-making abilities. To cultivate such competence, educational curricula must go beyond conventional theoretical instruction and systematically integrate buildability principles into course structures, learning outcomes, teaching methods, and assessment strategies. A well-designed curriculum not only imparts foundational knowledge but also develops professional judgment, interdisciplinary collaboration, problem-solving skills, and technological proficiency, all of which are essential for addressing the complexities of contemporary construction projects.

At its core, curriculum design for buildability competence should be intentionally structured around learning outcomes that reflect industry expectations and project realities. Traditionally, construction education has focused on the acquisition of technical knowledge in areas such as structural design, construction materials, and project management. While these are fundamental, a buildability-focused curriculum emphasises the practical application of this knowledge. For instance, students should be able to evaluate how a design decision affects construction sequencing, labour deployment, resource utilisation, and cost efficiency. Learning outcomes should explicitly target competencies such as constructability analysis, value engineering, risk assessment, and decision-making under constraints of time, cost, quality, and safety. By defining clear outcomes, curriculum designers create a roadmap that ensures students progressively develop the skills and judgment needed to enhance buildability.

A central feature of effective curriculum design is the adoption of an integrated and interdisciplinary approach. Construction projects inherently involve multiple disciplines, including architecture, civil and structural engineering, mechanical and electrical engineering, and construction management. Historically, educational programmes have tended to silo these disciplines, teaching them independently with minimal interaction. However, buildability challenges frequently arise at the interfaces between disciplines, where design intent, construction methods, and operational constraints intersect. Integrating disciplines within the curriculum allows students to understand the interdependencies that influence project deliverables. For example, a module that combines structural design with construction management principles enables students to appreciate how structural detailing decisions affect site productivity, material handling, and sequencing. Similarly, interdisciplinary studio projects or collaborative workshops simulate real-world scenarios where architects, engineers, and contractors must work together to resolve conflicts and optimise buildable designs. Such integrated learning experiences foster systems thinking and encourage students to view projects holistically, rather than as fragmented components.

In addition to interdisciplinary integration, a buildability-oriented curriculum must emphasise applied learning and experiential engagement. Theoretical knowledge alone is insufficient for developing the judgment required to assess construction feasibility. Students must actively apply concepts to realistic scenarios through problem-based learning, case studies, and project-based assignments. Constructability analysis exercises, for instance, can challenge students to critically evaluate design proposals for potential on-site challenges, inefficiencies, or risks. Value engineering tasks encourage them to identify cost-effective alternatives that maintain or improve performance, while risk assessment exercises develop their ability to anticipate and mitigate construction risks. These applied learning strategies bridge the gap between theory and practice, ensuring that students are equipped to make informed decisions that improve buildability.

Experiential learning opportunities such as site visits, internships, and industry-led workshops are particularly valuable for contextualising theoretical concepts. Exposure to actual construction sites allows students to observe sequencing, logistics, safety management, and construction methodologies in practice. Internships with construction firms or collaborative projects with industry partners further enable students to apply classroom learning to complex, real-world problems. Such experiences also cultivate professional skills like communication, teamwork, and adaptability, which are critical for navigating the challenges of multidisciplinary project environments. Reflective learning components, including journals, reports, or post-project evaluations, reinforce knowledge retention and encourage critical self-assessment.

The integration of digital and technological competencies is another essential aspect of curriculum design for buildability. The construction industry has increasingly adopted digital tools and platforms, such as BIM, digital collaboration environments, and construction simulation software. Proficiency in these technologies allows students to visualise construction sequences, detect design clashes, simulate workflows, and evaluate alternative construction approaches. For example, 4D BIM models enable students to analyse time-sequenced construction processes, assess resource allocation, and identify potential inefficiencies before they occur on-site. Simulation tools can be used to optimise site layouts, crane operations, or material delivery schedules, providing practical insights that directly enhance buildability. Embedding digital competencies in the curriculum ensures that graduates are not only technically capable but also able to leverage technology to improve collaboration, coordination, and project performance.

A comprehensive curriculum should also incorporate communication and collaboration skills, given their centrality to buildability. Buildable designs are rarely the product of individual effort; they result from coordinated input across multiple stakeholders. Curriculum design should therefore include collaborative projects, peer assessments, and interdisciplinary teamwork exercises. Students must learn to negotiate trade-offs, communicate technical information clearly, and reconcile differing priorities, whether related to aesthetics, cost, safety, or constructability. Developing these soft skills prepares students to function effectively in professional environments where clear communication and shared responsibility are vital for minimising design errors, reducing rework, and enhancing overall project efficiency.

Furthermore, curriculum design should address regulatory, environmental, and contextual awareness. Understanding building codes, safety standards, zoning regulations, and environmental constraints is critical for assessing constructability and ensuring compliance. Similarly, awareness of local construction practices, material availability, labour skills, and logistical limitations allows students to propose solutions that are feasible in real-world contexts. Curriculum content should include modules on construction law, occupational health and safety, sustainable construction practices, and project logistics, providing students with the knowledge necessary to apply buildability principles responsibly and practically.

To maintain relevance and effectiveness, curriculum design for buildability competence should be dynamic and responsive to industry evolution. Construction methods, materials, technologies, and regulatory frameworks continually evolve, and curricula must reflect these changes. Regular consultation with industry stakeholders, alumni, and professional bodies ensures that learning content remains aligned with current professional practice. Feedback loops, such as post-graduation surveys or employer evaluations, provide valuable insights into the effectiveness of curriculum elements in preparing graduates for buildability-focused roles. Additionally, incorporating opportunities for continuous professional development within the curriculum encourages lifelong learning, enabling students to keep pace with emerging trends and innovations.

Finally, assessment strategies play a key role in reinforcing buildability competence. Traditional examinations that focus solely on theoretical knowledge are insufficient for evaluating a student's ability to apply concepts in practice. Assessments should include project work, constructability reviews, simulations, collaborative assignments, and reflective reports that measure both technical proficiency and decision-making skills. Competency-based assessment approaches provide opportunities for students to demonstrate their ability to integrate knowledge, analyse alternatives, communicate effectively, and implement solutions that enhance buildability.

In summary, curriculum design for buildability competence must be multifaceted, integrating theoretical knowledge, practical application, digital proficiency, interdisciplinary collaboration, regulatory awareness, and industry engagement. By adopting an integrated and applied approach, curricula can develop graduates who are capable of evaluating designs from a construction-oriented perspective, identifying and resolving constructability challenges, and contributing to more efficient, safe, and high-quality project delivery. Through careful alignment of learning outcomes, teaching methods, experiential learning opportunities, technological integration, and assessment strategies, educational institutions can cultivate professionals equipped to meet the complex demands of contemporary construction practice and drive improvements in buildability across the industry.

To visually summarise the integration of curriculum components in developing buildability competence, Figure 8.1 presents a holistic framework that illustrates the interconnected elements of construction education. The diagram highlights how core technical knowledge, interdisciplinary collaboration, problem-solving and analysis, digital and technological training, experiential learning, and industry and contextual awareness collectively contribute to the development of skilled, industry-ready graduates capable of enhancing buildability in real-world projects. This framework emphasises that effective curriculum design is not linear but integrative, ensuring that students acquire the knowledge, skills, and practical experience necessary to apply buildability principles throughout the project lifecycle.

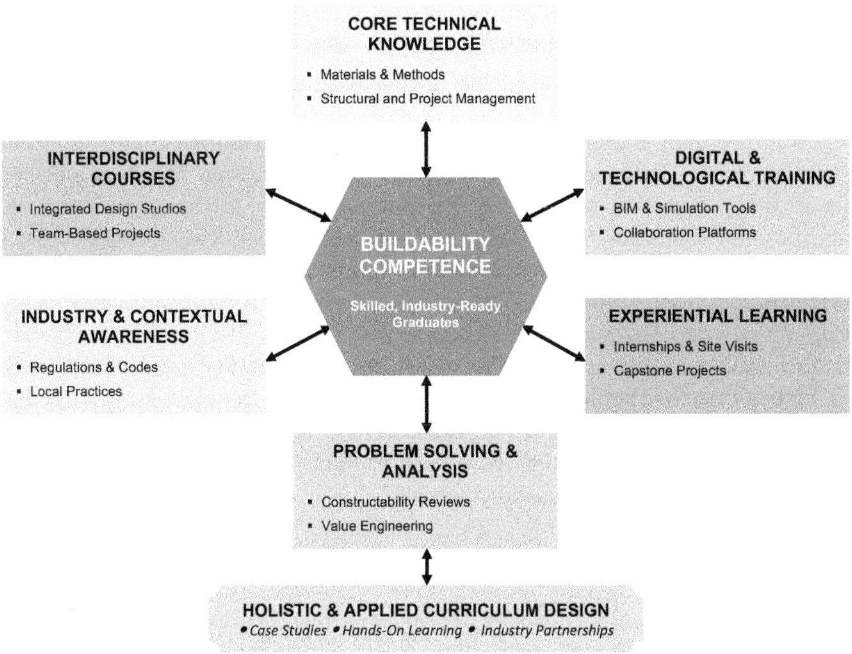

Figure 8.1 Integrated curriculum framework for developing buildability competence in construction education.

8.3 Pedagogical Approaches to Teaching Buildability in Construction Education

Teaching buildability in construction education requires pedagogical strategies that extend well beyond traditional lecture-based instruction. Buildability is a multidimensional competency that encompasses technical knowledge, practical problem-solving, critical thinking, interdisciplinary coordination, and the ability to make informed decisions under real-world constraints. Therefore, effective teaching of buildability necessitates approaches that actively engage students in both theoretical understanding and applied learning, equipping them with the skills and judgment necessary to anticipate and address constructability challenges throughout the project lifecycle. These approaches are discussed in the following subsections.

8.3.1 Active Learning and Problem-Based Approaches

Active learning methods are central to developing buildability competence. Unlike passive learning, active learning encourages students to engage directly with realistic problems, fostering deeper understanding and practical skills. Project-based learning (PBL) is particularly effective, as it immerses students in simulated construction projects where they must apply design, engineering, and construction principles to achieve feasible and efficient solutions. For instance, students can be tasked with evaluating the constructability of a proposed structural design, identifying sequencing issues, resource constraints, and potential safety hazards, and recommending solutions that optimise efficiency while maintaining quality and compliance.

PBL and scenario-based exercises further enhance critical thinking by presenting students with complex, ambiguous challenges that mirror real-world construction issues. For example, a scenario may require students to assess a design for an urban high-rise, considering material availability, labour constraints, site logistics, environmental factors, and regulatory compliance. This approach fosters decision-making under realistic constraints, teaching students to weigh trade-offs between cost, time, safety, quality, and sustainability. Additionally, case study analysis provides opportunities to examine past projects, learn from construction failures and successes, and internalise principles of buildability in diverse contexts.

Through active learning, students develop not only technical knowledge but also transferable skills such as problem-solving, analytical reasoning, and the ability to propose practical, implementable solutions, core competencies that are directly applicable to professional practice.

8.3.2 Collaborative and Interdisciplinary Learning

Buildability is inherently collaborative. Many challenges arise at the intersections between disciplines, such as architectural design, structural engineering, mechanical and electrical systems, and construction management. Consequently, pedagogical approaches must cultivate the ability to work effectively in interdisciplinary teams. Collaborative learning, which includes group projects, workshops, and studio-based exercises, enables students to experience real-world team dynamics, including negotiation, coordination, and conflict resolution.

For example, an interdisciplinary design studio may pair architecture students with civil engineering and construction management students. Together, they must produce a design that is aesthetically and functionally sound while remaining feasible to construct. Throughout this process, students learn to communicate across technical domains, reconcile differing priorities, and develop integrated solutions that balance aesthetics, performance, cost, and constructability. Such collaborative exercises mirror professional environments and help students understand that successful buildability outcomes depend on teamwork and shared responsibility.

Peer evaluation and reflective group discussions further reinforce collaborative learning. Students assess each other's contributions, reflect on team decision-making processes, and identify areas for improvement. These activities not only strengthen technical understanding but also develop leadership, communication, and interpersonal skills, competencies that are critical for promoting buildability in real-world construction projects.

8.3.3 Experiential and Industry-Linked Learning

Experiential learning is a cornerstone of buildability education, as it allows students to bridge the gap between theoretical knowledge and practical application. Exposure to live construction environments, through internships, site visits, and industry-led workshops, provides students with direct insight into construction processes, sequencing, logistics, and safety practices. Observing real-world projects helps students contextualise classroom knowledge, understand practical constraints, and appreciate the complexities involved in delivering buildable designs.

Capstone projects and industry-sponsored assignments offer further opportunities for experiential learning. For instance, students may be tasked with preparing a construction plan for a community building in collaboration with a local contractor. These projects often integrate digital tools, such as BIM, 4D simulations, and project management software, enabling students to analyse construction sequences, identify clashes, simulate workflows, and propose optimised solutions. Through these experiences, students develop practical judgment, problem-solving capabilities, and the confidence to apply buildability principles in real-world contexts.

Experiential learning also strengthens professional networks, as students interact directly with practitioners and industry mentors. Such connections provide insight into industry expectations, emerging trends, and the realities of construction project delivery, further enhancing students' readiness to contribute to buildable, efficient, and safe projects upon graduation.

8.3.4 Integration of Digital and Technological Tools

The integration of digital technologies into pedagogical approaches is increasingly vital for teaching buildability. BIM, digital collaboration platforms, construction simulation software, and visualisation tools provide students with opportunities to interact with complex construction systems in a controlled environment. BIM, for example, allows students to create detailed 3D models that incorporate architectural; structural; and mechanical, electrical, and plumbing components. These models can be analysed for potential clashes, sequencing inefficiencies, or constructability challenges, enabling iterative learning and informed decision-making.

4D simulations extend this capability by integrating time-based sequencing, allowing students to visualise construction processes over time. This enables assessment of scheduling conflicts, resource allocation, and logistical constraints, fostering an understanding of how construction decisions affect overall buildability. Similarly, digital collaboration platforms facilitate communication, coordination, and version control among student teams, simulating professional workflows and reinforcing interdisciplinary collaboration.

Technological proficiency in these tools also prepares students for the expectations of modern construction practice, where digital competencies are increasingly required for effective project delivery. By embedding these technologies within pedagogical strategies, educators equip students with the technical and analytical skills necessary to optimise design and construction processes, reduce errors, and improve overall buildability outcomes.

8.3.5 Reflective and Competency-Based Learning

Reflective practice is an essential component of pedagogical approaches for buildability. Encouraging students to critically evaluate their decisions, problem-solving approaches, and project outcomes fosters deeper understanding and continuous improvement. Reflection can be integrated through learning journals, post-project reviews, and structured self-assessment exercises. By considering what worked, what did not, and why, students develop metacognitive skills that enhance professional judgment and the ability to anticipate constructability challenges in future projects.

Competency-based assessment methods complement reflective learning by evaluating students' ability to apply knowledge and skills in practical contexts. Assessment can include project deliverables, constructability reviews, simulation exercises, digital model analysis, teamwork performance, and reflective reports. Such approaches move beyond rote memorisation, measuring not only what students know but also what they can do, ensuring that graduates demonstrate the practical competencies necessary to enhance buildability in professional practice.

8.3.6 Combining Pedagogical Approaches for Maximum Impact

A robust buildability-focused curriculum combines multiple pedagogical approaches to reinforce learning outcomes. Active learning develops critical thinking and problem-solving skills, collaborative exercises foster teamwork and interdisciplinary understanding, experiential learning provides real-world context, digital tools enable technical and analytical proficiency, and reflective practice ensures continuous improvement. When integrated effectively, these approaches create a dynamic, engaging, and holistic learning environment where students can internalise and apply buildability principles.

For example, a typical pedagogical sequence may involve students first exploring theoretical principles of constructability through lectures and case studies. This is followed by a collaborative design studio, where interdisciplinary teams create buildable solutions using BIM and simulation software. Students may then participate in a site visit or internship to observe real-world application, followed by reflective exercises and assessments to consolidate learning. By connecting theory, practice, collaboration, and technology, this approach ensures that students develop the knowledge, skills, and judgment required to deliver projects that are efficient, safe, cost-effective, and constructible.

In summary, pedagogical approaches to teaching buildability in construction education must be multifaceted, experiential, and integrative. Effective strategies combine active, problem-based learning with collaborative interdisciplinary projects, industry engagement, digital technology integration, and reflective practice. By engaging students in authentic, contextually rich learning experiences, educators can cultivate professionals who are capable of evaluating designs from a construction-oriented perspective, anticipating and resolving constructability challenges, and contributing to high-quality, efficient, and safe project delivery. The adoption of these pedagogical strategies ensures that construction graduates are not only technically competent but also equipped with the critical thinking, collaboration, and decision-making skills necessary to enhance buildability in

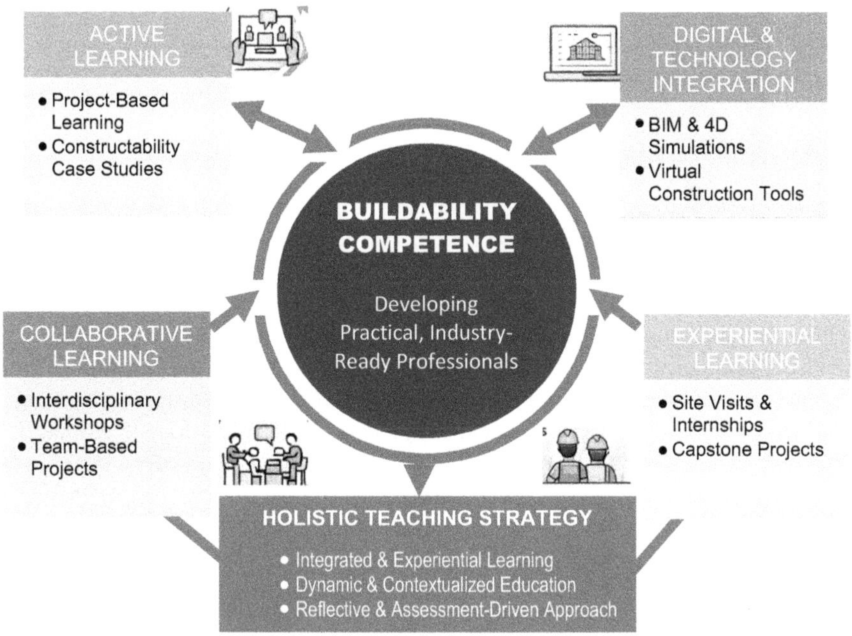

Figure 8.2 Framework of pedagogical approaches to teaching buildability in construction education.

contemporary construction practice. Figure 8.2 illustrates a holistic pedagogical framework for teaching buildability in construction education. It highlights the integration of active learning, collaborative interdisciplinary exercises, experiential industry engagement, and digital technology use, all underpinned by reflective and assessment-driven strategies. This framework demonstrates how diverse teaching approaches work together to equip students with the knowledge, skills, and practical judgment required to enhance constructability in real-world projects.

8.4 Industry Partnerships and Experiential Learning

In modern construction education, the integration of industry partnerships and experiential learning has become indispensable for cultivating buildability competence among students. Buildability is a multidimensional concept, encompassing not only technical knowledge of construction methods, materials, and design principles but also practical judgment, problem-solving, interdisciplinary collaboration, and decision-making under real-world constraints. While theoretical instruction provides the foundational knowledge, true competence in buildability requires exposure to authentic construction environments where students can observe, analyse, and participate in processes that influence constructability outcomes. Industry engagement and structured experiential learning bridge this critical gap between academic instruction and professional practice, enabling students to develop the technical and professional skills necessary to deliver efficient, safe, and high-quality construction projects.

8.4.1 The Role of Industry Partnerships

Industry partnerships are a central mechanism through which construction education aligns with professional practice. Partnerships with construction firms, architects, engineers, consultants, and material suppliers provide students with access to the latest technologies, methods, and project management practices. These collaborations allow institutions to co-develop curricula, ensuring that learning outcomes reflect current industry requirements and expectations. For example, industry partners can provide real project briefs, case studies, or design challenges that incorporate practical constraints such as site logistics, material availability, labour skill levels, regulatory requirements, and safety considerations. By working on these authentic scenarios, students learn to apply theoretical knowledge to practical problems while understanding the complexities and limitations of real-world projects.

Industry partnerships also facilitate mentorship programmes, guest lectures, and workshops. Practitioners bring valuable insights into professional decision-making processes, constructability challenges, and best practices. Through these engagements, students gain a nuanced understanding of how design decisions impact construction efficiency, cost, and safety, fostering the development of professional judgment critical for evaluating buildability. Additionally, industry partnerships often provide networking opportunities, helping students establish professional relationships that support career development and ongoing learning beyond formal education.

8.4.2 Experiential Learning Approaches

Experiential learning, often described as "learning by doing," is a pedagogical approach that immerses students in authentic construction environments, encouraging them to apply knowledge, observe processes, and make decisions in real-world contexts. Experiential learning is particularly effective for enhancing buildability competence because it enables students to witness and engage with the dynamic interplay of design, construction methods, resources, and site conditions.

Internships and apprenticeships are among the most common forms of experiential learning in construction education. During internships, students are exposed to practical challenges such as scheduling conflicts, sequencing of construction activities, material deliveries, and coordination of subcontractors. They may observe how design intent is translated into executable construction plans or participate in constructability assessments that identify potential risks, inefficiencies, or safety concerns. Through this exposure, students develop a practical understanding of how theoretical principles, such as structural integrity, site logistics, and cost optimisation, are applied and adapted in real-world situations.

Capstone projects and live industry collaborations provide another avenue for experiential learning. These projects typically involve collaboration with a construction firm or design consultancy, where students are required to integrate technical knowledge, interdisciplinary coordination, and digital tools to develop buildable solutions. For example, a capstone project may involve using BIM to simulate construction sequences, identify potential clashes, and optimise workflows. Students may also conduct constructability reviews, perform value engineering exercises, or analyse project risks to propose improvements. These experiences cultivate critical thinking, problem-solving skills, and professional judgment, while reinforcing the practical application of theoretical knowledge.

Site visits and field exposure are additional experiential learning strategies that enhance students' understanding of buildability. Observing live construction sites allows students to see the real-time implications of design decisions, resource allocation, and site management practices. It also exposes them to safety management systems, quality control procedures, and environmental compliance requirements. These visits complement classroom instruction by providing tangible, context-rich examples of how buildability principles are applied, helping students internalise lessons that are difficult to convey through lectures alone.

8.4.3 Integration of Digital Tools in Experiential Learning

Digital and technological competencies are increasingly important in both classroom and experiential learning environments. Tools such as BIM, 4D construction simulations, and project management software enable students to visualise construction sequences, identify design conflicts, and test alternative construction strategies. For instance, during a site-based project, students can compare the digital model with actual site conditions, noting deviations, constraints, and constructability challenges. Digital tools allow iterative learning, where students can refine their solutions based on observed outcomes or simulation results. Moreover, digital platforms support collaboration among student teams and industry mentors, providing a virtual environment that complements on-site experiential learning.

8.4.4 Benefits of Industry Partnerships and Experiential Learning

The combination of industry partnerships and experiential learning offers multiple benefits for developing buildability competence:

1. Contextual Understanding: Students gain first-hand knowledge of regulatory requirements, site conditions, labour skills, material availability, supply chain constraints, and environmental factors, which are essential for practical decision-making.
2. Application of Theoretical Knowledge: Experiential learning enables students to apply concepts learned in the classroom, reinforcing learning outcomes and promoting deeper understanding.

3 Development of Professional Skills: Exposure to collaborative projects and real-world challenges develops communication, teamwork, leadership, and negotiation skills, competencies that are essential for coordinating multiple stakeholders to enhance buildability.
4 Integration of Technology: Students learn to leverage digital tools in practical contexts, enhancing their ability to evaluate construction sequences, detect design clashes, and optimise workflows.
5 Industry-Ready Graduates: Continuous engagement with professionals ensures that graduates are equipped to meet the expectations of employers, contributing effectively to project planning, constructability analysis, and efficient delivery of high-quality construction projects.

8.4.5 Best Practices for Implementation

To maximise the impact of industry partnerships and experiential learning, educational institutions should adopt a structured and systematic approach:

1 Formal Industry Agreements: Establish partnerships with construction firms, consultants, and professional associations to provide structured internships, site visits, and live project collaborations.
2 Curriculum Integration: Embed experiential learning projects and case studies into course modules with clear learning objectives related to buildability. This ensures that industry experiences are directly linked to academic outcomes.
3 Faculty-Industry Collaboration: Encourage educators and industry professionals to co-design project briefs, mentorship programmes, and assessment criteria, ensuring alignment with current industry practices and technological trends.
4 Reflective Practice: Incorporate structured reflection exercises, reports, and presentations, allowing students to evaluate their learning, decision-making processes, and professional development.
5 Digital and Remote Engagement: Use virtual site tours, BIM-based exercises, and simulation platforms when on-site access is limited, ensuring continuity of experiential learning.

In summary, industry partnerships and experiential learning are fundamental to the development of buildability competence in construction education. By providing students with exposure to real-world construction environments, access to current industry knowledge, and opportunities for applied learning, these strategies bridge the gap between theory and practice. Experiential learning enables students to internalise constructability principles, understand the practical constraints of projects, and develop critical professional skills such as teamwork, problem-solving, and decision-making. Simultaneously, industry partnerships ensure curricula remain relevant, reflective of current practice, and aligned with professional expectations. When effectively implemented, these approaches produce graduates who are not only technically proficient but also capable of evaluating, optimising, and delivering buildable construction projects, contributing to safer, more efficient,

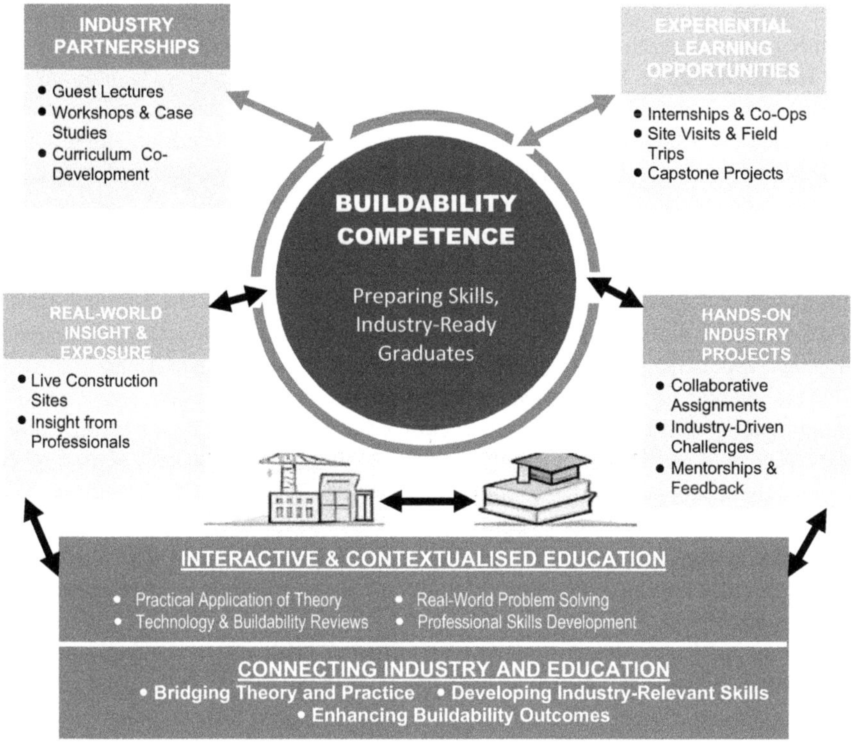

Figure 8.3 Industry partnerships and experiential learning framework for developing buildability competence in construction education.

and higher-quality outcomes in the construction industry. Figure 8.3 illustrates how industry partnerships and experiential learning collaboratively enhance buildability competence in construction education. The framework highlights key components, including internships, site visits, capstone projects, mentorship, and industry-driven assignments, as well as the integration of digital tools and professional guidance. Together, these elements provide students with interactive, context-rich experiences that bridge the gap between theory and practice, strengthen practical skills, and prepare graduates to contribute effectively to constructible, efficient, and high-quality construction projects.

8.5 Chapter Conclusion

This chapter has emphasised that enhancing buildability in construction is fundamentally an educational and capacity-building endeavour. While technical tools, procurement strategies, and regulatory frameworks are important, the ability to design and deliver buildable projects ultimately depends on the competencies of professionals across the project lifecycle. Developing these competencies requires a deliberate and structured approach to education and training, one that aligns academic preparation with the practical realities of construction practice.

The discussion on key competencies highlighted that buildability is multidimensional. Foundational knowledge in construction methods, materials, and structural principles must be complemented by design-construction integration skills, interdisciplinary communication, and critical thinking. The ability to anticipate site constraints, coordinate across disciplines, and resolve constructability challenges early in the design phase is central to achieving efficient and practical project outcomes. Increasingly, digital and technological competence, particularly in tools such as BIM, simulation platforms, and data-driven planning systems, has become essential. However, technical proficiency alone is insufficient without contextual awareness and practical experience that ground decision-making in real-world constraints.

This chapter further demonstrated that curriculum design plays a crucial role in embedding buildability into construction education. Rather than treating buildability as an isolated topic, it should be integrated across courses in design, project management, construction technology, and digital systems. A coherent curriculum must progressively develop competencies, linking theoretical knowledge to applied practice. This integration ensures that learners understand not only how to design but also how to design for construction efficiency, safety, cost-effectiveness, and sustainability.

Pedagogical approaches were also identified as critical to achieving meaningful learning outcomes. Active learning and problem-based approaches encourage students to engage with real construction challenges, fostering analytical and decision-making skills. Collaborative and interdisciplinary learning mirrors the team-based nature of construction projects, preparing learners for professional practice. Experiential and industry-linked learning, including site visits, internships, and live projects, bridges the gap between classroom instruction and field realities. The integration of digital tools within these approaches enhances technical fluency, while reflective and competency-based learning supports continuous professional development. Importantly, this chapter underscored that combining these pedagogical strategies yields the greatest impact, creating a dynamic and practice-oriented learning environment.

Industry partnerships emerged as a cornerstone for strengthening buildability education. Collaboration between academic institutions and industry stakeholders ensures that curricula remain relevant, responsive, and aligned with evolving industry demands. Experiential learning opportunities supported by industry, such as mentorship programmes, joint research initiatives, and real-world project collaborations, provide students with exposure to authentic construction processes and decision-making contexts. The strategic integration of digital tools in these partnerships further enhances experiential learning, allowing students to engage with contemporary practices and technologies.

In conclusion, advancing buildability in construction requires a systemic commitment to education and training reform. By cultivating integrated competencies, designing coherent curricula, adopting innovative pedagogical methods, and fostering strong industry partnerships, educational institutions can prepare professionals who are capable of delivering practical, efficient, and sustainable construction solutions. Ultimately, embedding buildability into construction education strengthens not only individual competence but also the overall performance and resilience of the construction sector.

Bibliography

Agyekum, K., Sackey, K.N.Y.S., Addoh, F.E., Pittri, H., Sosu, J., & Danso, F.O. (2025). Key competencies of built environment professionals for achieving net-zero carbon emissions in the Ghanaian construction industry. *Buildings*, 15(10), 1750. https://doi.org/10.3390/buildings15101750

Alnajjar, O., Atencio, E., & Turmo, J. (2025). Framework for optimizing the construction process: The integration of lean construction, building information modeling (BIM), and emerging technologies. *Applied Sciences*, 15(13), 7253. https://doi.org/10.3390/app15137253

Alsmadi, H., Kandasamy, G., Al Kafri, A., & Zahirah, K.F. (2024). Empowering computing students through multidisciplinary project based learning (PBL): Creating meaningful differences in the real world. *Social Sciences & Humanities Open*, 10, 101180. https://doi.org/10.1016/j.ssaho.2024.101180.

Atkinson, A.R., & Westall, R. (2010). The relationship between integrated design and construction and safety on construction projects. *Construction Management and Economics*, 28(9), 1007–1017. https://doi.org/10.1080/01446193.2010.504214

Awe, M., Malhi, A., Budka, M., Mavengere, N., & Dave, B. (2025). Towards 4D BIM: A systematic literature review on challenges, strategies and tools in leveraging AI with BIM. *Buildings,* 15, 1072. https://doi.org/10.3390/buildings15071072

Chakkol, M., Selviaridis, K., & Finne, M. (2018), The governance of collaboration in complex projects. *International Journal of Operations & Production Management*, 38(4), 997–1019. https://doi.org/10.1108/IJOPM-11-2017-0717

Cherbonnier, A., Hémon, B., Michinov, N., Jamet, E., & Michinov, E. (2024). Collaborative skills training using digital tools: A systematic literature review. *International Journal of Human–Computer Interaction*. https://doi.org/10.1080/10447318.2024.2348227

Dixit, M.K., Venkatraj, V., Pariafsai, F., & Bullen, J. (2022). Site logistics factors impacting resource use on construction sites: A Delphi study. *Frontiers in Built Environment*, 8, 858135. https://doi.org/10.3389/fbuil.2022.858135

Dosumu, O., & Aigbavboa, C. (2018). Concept of sustainability and sustainable design and construction. *In book: Sustainable design and construction in Africa* (1st ed.). Routledge.

Faraji, A., Rashidi, M., Meydani Haji Agha, T., Rahnamayiezekavat, P., & Samali, B. (2022). Quality management framework for housing construction in a design-build project delivery system: A BIM-UAV approach. *Buildings*, 12(5), 554. https://doi.org/10.3390/buildings12050554

Fu, J., Hutson, B., Thierer, K., Zhang, X., & Taylor, J. (2025). Experiential learning assessment: A dive into the literature. *Intersection: A Journal at the Intersection of Assessment and Learning*, 6(1), 3–20. https://doi.org/10.61669/001c.131785

Gangolells, M., Casals, M., Forcada, N., Roca, X., & Fuertes, A. (2010). Mitigating construction safety risks using prevention through design. *Journal of Safety Research*, 41(2), 107–122. https://doi.org/10.1016/j.jsr.2009.10.007.

Górecki, J., & Núñez-Cacho, P. (2022). Decision-making problems in construction projects executed under the principles of sustainable development—Bridge construction case. *Applied Sciences*, 12(12), 6132. https://doi.org/10.3390/app12126132

Görsch, C., Seppänen, O., Peltokorpi, A., & Lavikka, R. (2024). Task planning and control in construction: Revealing workers as early and late planners. *Construction Management and Economics*, 42(5), 431–450. https://doi.org/10.1080/01446193.2023.2270080

Hashempour, N., Zadeh, P.A., & Staub-French, S. (2024). Understanding the integration of building energy modeling into the building design process: Insights from two collaborative construction projects. *Buildings*, 14(11), 3379. https://doi.org/10.3390/buildings14113379

Kannimuthu, M. (2025). Effective site logistics and spatial layout planning for construction efficiency. In: Bajaj, D. (Eds.) *Handbook of construction project management.* Springer. https://doi.org/10.1007/978-981-96-7631-6_16

Khan, A.M., Alaloul, W.S., Musarat, M.A., & Fayyaz, A.M. (2025). Optimizing sustainable alternatives in value engineering decision-making through BIM-integrated plugin automation for buildings. *Ain Shams Engineering Journal*, 16(6), 103373. https://doi.org/10.1016/j.asej.2025.103373.

Kolajo, Y. (2025). Advancing pedagogical excellence through reflective teaching practice and adaptation. *Reflective Practice*, 26(6), 832–847. https://doi.org/10.1080/14623943.2025.2504143

Lai, H., & Chang, T.Y.P. (2019). BIM-based platform for collaborative building design and project management. *Journal of Computing in Civil Engineering*, 33(3). https://doi.org/10.1061/(ASCE)CP.1943-5487.0000830

Larsson, J., & Larsson, L. (2020). Integration, application and importance of collaboration in sustainable project management. *Sustainability*, 12(2), 585. https://doi.org/10.3390/su12020585

Lin, S.S., Shen, S.L., Zhou, A., & Chen, X.S. (2025). Smart techniques promoting sustainability in construction engineering and management. *Engineering*, 45, 262–282. https://doi.org/10.1016/j.eng.2024.08.023.

Mei, T., Guo, Z., Li, P., Fang, K., & Zhong, S. (2022). Influence of integrated project delivery principles on project performance in China: An SEM-based approach. *Sustainability*, 14(8), 4381. https://doi.org/10.3390/su14084381

Meierdirk, C. (2016). Is reflective practice an essential component of becoming a professional teacher? *Reflective Practice*, 17(3), 369–378. https://doi.org/10.1080/14623943.2016.1169169

Nechyporchuk, Y., & Bašková, R. (2022). Tools of 4D modeling of construction projects. *IOP Conference Series Materials Science and Engineering*, 1252(1), 012078. https://doi.org/10.1088/1757-899X/1252/1/012078

Nwulu, E.O., Elete, T.Y., Omomo, K.O., Akano, O.A., & Erhueh, O.V. (2023). The importance of interdisciplinary collaboration for successful engineering project completions: A strategic framework. *World Journal of Engineering and Technology Research*, 02(03), 001–016. https://doi.org/10.53346/wjetr.2023.2.3.0048

Osuizugbo, I.C. (2020). The need for and benefit of buildability analysis: Nigeria as a case study. *Journal of Engineering, Design and Technology*, 19(5), 1207–1230. https://doi.org/10.1108/JEDT-08-2020-0338

Osuizugbo, I.C. (2025). Buildability-led procurement: A new model for future construction projects delivery. *Construction Innovation: Information, Process, Management*. https://doi.org/10.1108/CI-10-2023-0272

Osuizugbo, I.C., Okolie, K.C., Oshodi, O.S., & Oyeyipo, O.O. (2023). Buildability in the construction industry: A systematic review. *Construction Innovation: Information Process Management*, 23(5), 1300–1322. https://doi.org/10.1108/CI-05-2022-0112

Pandit, A. (2023). The impact of value engineering on pre-construction cost control: Striking the balance between quality and budget. *International Scientific Journal of Engineering and Management*, 02(11), 1–8. https://doi.org/10.55041/ISJEM01299

Patil, K.R., Ayer, S.K., McCord, K.H., Perry, L.A., Wu, W., London, J.S., & Kline, A.R. (2025). Characteristics of authentic construction learning experiences to enable accurate consideration of cost-effective alternatives. *Buildings*, 15(14), 2446. https://doi.org/10.3390/buildings15142446

Ponomarioviené, J., Jakavonytė-Staškuviené, D., & Torterat, F. (2025). Implementing competency-based education through the personalized monitoring of primary students' Progress and Assessment. *Education Sciences*, 15(2), 252. https://doi.org/10.3390/educsci15020252

Puolitaival, T., Davies, K., & Kähkönen, K. (2019). Digital technologies and related competences in construction management in the era of fast-paced digitalisation. In: *Conference: CIB world building Congress* 2019, Hong Kong.

Puwakkatiya Lokuge, I.W., Kulatunga, U., Gajendran, T., & Beard, C. (2025). Technological advancements and their influence on buildability in construction projects: A phenomenological analysis. *Construction Innovation: Information Process Management*,. https://doi.org/10.1108/CI-04-2025-0153

Rakha, A.H. (2025). Promoting online teaching through active learning strategies: Applications and innovations. *Frontiers in Education*, 10, 1546208. https://doi.org/10.3389/feduc.2025.1546208

Rehman, N., Huang, X., Mahmood, A., AlGerafi, M.A.M., & Javed, S. (2024). Project-based learning as a catalyst for 21st-century skills and student engagement in the math classroom. *Heliyon*, 10(23), e39988. https://doi.org/10.1016/j.heliyon.2024.e39988.

Rezai, A., Ahmadi, R., Ashkani, P., & Hosseini, G.H. (2025). Implementing active learning approach to promote motivation, reduce anxiety, and shape positive attitudes: A case study of EFL learners. *Acta Psychologica*, 253, 104704. https://doi.org/10.1016/j.actpsy.2025.104704.

Russell, J.S., Hanna, A., Bank, L.C., & Shapira, A. (2007). Education in construction engineering and management built on tradition: Blueprint for tomorrow. *Journal of Construction Engineering and Management,* 133(9). https://ascelibrary.org/doi/10.1061/%28ASCE%290733-9364%282007%29133%3A9%28661%29

Sánchez-García, R., & Reyes-de-Cózar, S. (2025). Enhancing project-based learning: a framework for optimizing structural design and implementation—A systematic review with a sustainable focus. *Sustainability*, 17(11), 4978. https://doi.org/10.3390/su17114978

Sarvari, H., Edwards, D.J., Rillie, I., & Posillico, J.J. (2024). Building a safer future: Analysis of studies on safety I and safety II in the construction industry. *Safety Science*, 178, 106621. https://doi.org/10.1016/j.ssci.2024.106621

Shehadeh, A., Alshboul, O., Taamneh, M.M., Jaradat, A.Q., & Alomari, A.H. (2024). Enhanced clash detection in building information modeling: Leveraging modified extreme gradient boosting for predictive analytics. *Results in Engineering*, 24, 103439. https://doi.org/10.1016/j.rineng.2024.103439

Singh, J. (2025). Risk assessment and mitigation in construction building projects. *Journal of Construction and Building Materials Engineering*, 11(1), 14–23. https://doi.org/10.46610/JOCBME.2025.v011i01.002

Sulankivi, K., Tauriainen, M., & Kiviniemi, M. (2014). Safety aspect in constructability analysis with BIM. In: *Conference: CIB W099 international conference achieving sustainable construction health and safety*, Lund University, Lund, 2–3 June 2014, pp. 586–596.

Wang, K., Zhang, C., Guo, F., & Guo, S. (2022). Toward an efficient construction process: What drives BIM professionals to collaborate in BIM-enabled projects. *Journal of Management in Engineering*, 38(4). https://doi.org/10.1061/(ASCE)ME.1943-5479.0001056.

Waqar, A., Al Hajj, F., Danish, M., Khan, M.B., & Houda, M. (2025). Evaluating the influence of modern construction methods on sustainable construction success. *Cleaner and Responsible Consumption*, 19, 100327. https://doi.org/10.1016/j.clrc.2025.100327.

Wimalaratne, P.L.I., Kulatunga, U., & Gajendran, T. (2023). 'Buildability' in the digital age: A phenomenological discourse of industry practitioners' perceptions. *Buildings*, 13(11), 2870. https://doi.org/10.3390/buildings13112870

Wong, F.W.H., Lam, P.T.I, Chan, E.H.W., & Wong, F.K.W. (2006). Factors affecting buildability of building designs. *Canadian Journal of Civil Engineering*. 33(7): 795–806. https://doi.org/10.1139/l06-022

Yoou, S., Chang, T., & Chi, S. (2024). Developing an integrated construction safety management system for accident prevention. *Journal of Management in Engineering*, 40(6). https://doi.org/10.1061/JMENEA.MEENG-6074

Zamiri, M., & Esmaeili, A. (2024). Strategies, methods, and supports for developing skills within learning communities: A systematic review of the literature. *Administrative Sciences*, 14(9), 231. https://doi.org/10.3390/admsci14090231

Zhang, Q., & Ramli, M.F. (2025). A review of project-based learning and design education. *International Journal of Academic Research in Business and Social Sciences*, 15(1), 1464–1483. https://doi.org/10.6007/IJARBSS/v15-i1/24502

Zhang, J., Xie, H., & Li, H. (2019). Improvement of students problem-solving skills through project execution planning in civil engineering and construction management education. *Engineering, Construction and Architectural Management*, 26(7), 1437–1454. https://doi.org/10.1108/ECAM-08-2018-0321

Part IV

Sustainability and Future Pathways

9 Buildability and Sustainable Construction

9.1 Contributions of Buildability to Sustainability Futures

Buildability is increasingly understood as a decisive pathway through which sustainability ambitions are translated into feasible, efficient, and durable sustainable construction outcomes (Osuizugbo et al., 2023). While early buildability discourse centred on simplifying design to reduce construction difficulty (Hyde, 1995; Pheng Low, 2001; Crowther, 2002; Mbamali et al., 2005; Wong et al., 2006), recent research indicates that buildability now functions as a system-level sustainability enabler, influencing environmental performance (waste and embodied carbon) (Dasari et al., 2023; Caroli et al., 2022; Karimi et al., 2023; Laovisutthichai et al., 2022), economic performance (cost certainty and productivity) (Osuizugbo, 2025), and social outcomes (quality, safety, and disruption minimisation) across the project lifecycle (Osuizugbo et al., 2023).

9.1.1 Conceptual Evolution and Significance for Sustainability Futures

A core contribution of buildability to sustainability futures is its conceptual expansion from a narrow design characteristic to an integrated project performance construct. In a systematic review of extant literature, Osuizugbo et al. (2023) show that buildability literature has become increasingly multidimensional, encompassing design factors (simplicity, standardisation, modularity), process factors (planning, sequencing), organisational factors (communication, procurement), and technological factors (BIM and digital tools). This expanded framing is relevant for attaining sustainability futures because it aligns buildability with the governance, coordination, and decision structures that determine whether sustainability-oriented aspirations within construction project environments are practically deliverable.

As sustainability targets intensify through decarbonisation policies, circular economy (CE) strategies, and performance-based procurement, buildability has become a veritable mechanism for translating an ambitious design intent into a construction reality. In this sense, buildability not only improves efficiency but also protects sustainability-oriented value by reducing the likelihood that complex sustainability features fail during execution or become cost-prohibitive and are subsequently value-engineered out of the project. While the terms buildability

DOI: 10.1201/9781003747284-13

and constructability are frequently used interchangeably in practice, the literature reviewed here supports a broader, system-level interpretation of buildability that spans technical feasibility, organisational coordination, and lifecycle delivery conditions (Osuizugbo et al., 2023; Wimalaratne et al., 2021).

9.1.2 Waste Reduction, Rework Avoidance, and Embodied Carbon Mitigation

One of the most substantive contributions of buildability towards improving sustainability performance within the construction project environment lies in its capacity to minimise rework and material waste: two predominant sources of embodied carbon and resource inefficiency. When sustainability outcomes rely on highly precise installation processes (for example, airtightness layers, insulation continuity, and façade detailing), insufficient buildability can give rise to defects that necessitate demolition and reinstatement, thereby undermining both environmental and economic objectives. In such circumstances, construction rework generates avoidable consumption of additional materials, energy, labour, transport, and time, ultimately exacerbating a project's embodied carbon footprint and diminishing its overall resource efficiency (Mahamid, 2022; Love et al., 2018).

Singhaputtangkul et al. (2011) illustrate this interdependence in building envelopes, where they demonstrate that sustainability benefits such as improved thermal performance and energy efficiency can be compromised when design solutions create installation complexity or demand specialised methods that are not feasible within typical project delivery environments. The projected sustainability gains of an advanced building envelope may be lost through poor constructability, defects, and corrective work. Impliedly, the ability of buildability to support sustainability futures by acting as a risk-reduction mechanism against resource-intensive remediation can easily be discerned.

In industrialised and digital fabrication contexts, waste reduction is addressed more explicitly as a measurable performance outcome. Nerella et al. (2020) show that buildability assessment in 3D-printable concrete must extend beyond structural feasibility to include economic viability; trial-and-error printing increases waste and cost, limiting scalability and sustainability impact. Zhu et al. (2024) further demonstrate that predictive simulation (Peridynamics) can evaluate geometric stability before printing, enabling design refinements that reduce failed prints and material wastage. These studies collectively position buildability as a contributor to embodied carbon mitigation through reduced discard, improved first-time-right production, and better process predictability.

9.1.3 Early-Stage Decision-Making as a Sustainability Lever

Sustainability outcomes in construction projects are disproportionately shaped by early decisions: structural forms, envelope concepts, material strategies, and delivery methods chosen in concept and developed design constrain later possibilities (Singhaputtangkul et al., 2014). Accordingly, buildability reinforces sustainable futures by ensuring that early sustainability intentions remain achievable across all later stages of the built asset lifecycle.

Wimalaratne et al. (2023) argue that in the digital age, buildability is increasingly shaped by collaborative digital processes and practitioner experience rather than being purely a technical design attribute. This is important for sustainability because early coordination through BIM, integrated information, and model alignment can prevent later-stage conflicts that lead to redesign and rework. Buildability thus operates as a front-end control mechanism for sustainability risk.

This early-stage importance is reinforced by evidence that buildability can be co-optimised with sustainability. Mathern et al. (2022) show that multi-objective optimisation methods can incorporate both sustainability goals (such as reduced material consumption and environmental impact) and buildability requirements, enabling better early trade-off management. This approach supports sustainability futures by shifting decision-making from intuition-driven compromise to transparent, evidence-informed selection of design solutions that are both low-impact and realistically buildable.

9.1.4 Buildability as a Socio-Technical Pathway for Sustainable Delivery

A notable contemporary contribution is the recognition that buildability is socio-technical, shaped by people, processes, and tools in combination. Osuizugbo et al. (2023) highlight that buildability is applied differently across regions and project types, with inconsistent definitions limiting comparability. Wimalaratne et al. (2023) extend this by showing that practitioners' perceptions of buildability vary by role, experience, and project delivery method. This points to a sustainability-relevant reality: even strong technical solutions can fail if organisational culture, collaboration structures, or skill levels are insufficient to implement them.

From a sustainability futures perspective, buildability contributes to decarbonisation not merely through "better design," but through improved coordination and shared decision-making. Digital collaboration is one practical manifestation of this socio-technical alignment, as it can reduce fragmentation, strengthen early contractor involvement, and enable construction-informed sustainability decisions (Lesniewski & Berkebile, 2020). Conversely, fragmented workflows and low digital maturity can prevent sustainability-oriented design features from being executed as intended (Diaz Schery et al., 2025).

9.1.5 Lifecycle Performance: Durability, Maintainability, and Long-Term Value

Sustainability is not only about reduced construction-phase impacts but is also about long-term performance and reduced lifecycle cost. Buildability contributes to sustainability futures by increasing the likelihood that built assets are durable, serviceable, and maintainable.

Ogbeifun et al. (2023) frame buildability within a wider set of infrastructure performance criteria, emphasising the need to harmonise constructability with stability, serviceability, maintainability, cost, and functionality. This framing implies that buildability influences sustainability by ensuring that assets are delivered in a manner that supports long-term operational reliability and reduces maintenance burdens. Poor buildability performance can result in hidden lifecycle costs through premature defects, service disruptions, and difficult-to-maintain systems.

In housing contexts, this lifecycle contribution is particularly important because low-income and affordable housing requires solutions that are not only efficient to build but also resilient and maintainable under resource constraints. Studies assessing prefabrication in Southern Africa show that sustainability outcomes depend on contextual feasibility, supply chains, and skills availability (Moghayedi & Awuzie, 2023). Buildability is therefore essential for ensuring that sustainability strategies are durable, scalable, and equitable.

9.1.6 Buildability as a Precondition for Scalable Sustainability Innovation

A central sustainability futures question is how sustainable innovations move from pilots to mainstream adoption. Buildability contributes by improving the scalability of sustainable construction methods. If sustainable methods are too complex, skill-intensive, or uncertain in performance, they remain niche, and their sustainability benefits remain limited.

The additive manufacturing studies illustrate this clearly: buildability criteria and standardised tests support repeatability and comparability, which are preconditions for industrial scale-up (Nerella et al., 2020; Zhu et al., 2024). In optimisation research, buildability constraints ensure that sustainability improvements do not rely on idealised conditions but remain feasible in real delivery environments (Mathern et al., 2022). In envelope design, integrating buildability prevents sustainability-oriented solutions from being undermined by on-site complexity (Singhaputtangkul et al., 2011). Collectively, these studies position buildability as a bridge between sustainability innovation and practical adoption.

In summary, the literature indicates that buildability contributes to sustainability futures by reframing constructability as a system-level performance construct; reducing waste and rework that inflate embodied carbon; enabling early-stage optimisation and proactive coordination; strengthening socio-technical integration across roles and delivery models; improving lifecycle durability and maintainability; and supporting the scalability of sustainable innovations through repeatable assessment and feasible delivery pathways (Mathern et al., 2022; Nerella et al., 2020; Osuizugbo et al., 2023; Singhaputtangkul et al., 2011; Wimalaratne et al., 2023; Zhu et al., 2024).

9.1.7 Transitional Synthesis: From Contribution to Implementation

While Section 9.1 shows that buildability can strengthen sustainability outcomes through waste reduction, lifecycle optimisation, and improved coordination, these benefits do not materialise automatically. Evidence suggests that buildability's sustainability potential depends on how it is embedded within governance structures, delivery models, organisational cultures, and enabling technologies (Osuizugbo et al., 2023; Wimalaratne et al., 2023). Accordingly, this chapter now shifts from contribution to implementation. Section 9.2 identifies the strategic pathways through which buildability can be institutionalised as a practical mechanism for measurable sustainability performance.

9.2 Strategic Pathways for Integrating Buildability into Sustainable Construction Practices

Although buildability has been established as a significant contributor to sustainability futures, its impact depends on how effectively it is embedded within project systems. Extant literature consistently indicates that buildability does not automatically generate sustainability benefits; rather, it must be deliberately integrated into governance structures, design processes, technological frameworks, and organisational cultures.

A primary pathway is early-stage embedding. Sustainability performance is largely determined during conceptual and design development phases, when decisions regarding structural systems, material strategies, and delivery approaches remain flexible. Wimalaratne and Kulatunga (2025) conceptualise this cumulative effect as "buildability momentum," demonstrating that early integration of constructability considerations strengthens coordination and reduces downstream inefficiencies. Poor early decisions, by contrast, create negative momentum that becomes increasingly difficult and costly to correct. The implication is that buildability must be embedded at project inception rather than treated as a late-stage compliance check.

Early-stage integration is reinforced by decision-support technologies. Ballal and Sher (2003) illustrate that artificial neural networks can assist in selecting structural systems based on buildability criteria during preliminary design. By formalising expert knowledge into predictive tools, such approaches enable designers to evaluate feasibility before complexity becomes embedded in the project. When sustainability objectives are incorporated into these early evaluations, buildability functions as a safeguard against impractical or economically unsustainable design ambitions.

However, technical tools alone are insufficient. Socio-technical alignment is equally essential. Practitioner-focused research demonstrates that buildability outcomes are shaped by organisational culture, collaboration structures, and professional roles (Wimalaratne et al., 2023; PuwakkatiyaLokuge et al., 2025). Digital platforms and advanced modelling tools can enhance coordination, but only where teams possess the skills, trust, and institutional support required to use them effectively. In fragmented or adversarial project environments, sustainability-oriented design features may be diluted or abandoned due to late-stage constructability concerns. Consequently, integrating buildability into sustainable construction requires collaborative procurement models, early contractor involvement, and interdisciplinary design processes that promote shared ownership of constructability objectives.

Structured feedback mechanisms further strengthen integration. Henderson (2013) argues that many constructability failures arise from weak design–construction feedback loops, where experiential knowledge from site operations is not systematically captured and reintegrated into future design decisions. Strengthening these feedback loops through formal review cycles, knowledge management systems, and digital collaboration platforms enhances organisational learning and reduces recurring errors. From a sustainability perspective, this reduces rework, conserves resources, and improves delivery reliability over time.

Another strategic requirement is the development of measurable and standardised evaluation frameworks. Osuizugbo et al. (2023) highlight the conceptual fragmentation of buildability research and the limited availability of quantitative assessment methods. In digital fabrication contexts, Casagrande et al. (2020) demonstrate that buildability outcomes can vary significantly depending on testing procedures and process parameters, underscoring the importance of methodological consistency. Without measurable indicators, buildability remains difficult to benchmark or integrate into sustainability reporting systems. Establishing performance metrics, such as reductions in rework rates, material waste, and delivery time, would strengthen the empirical linkage between buildability and sustainability outcomes.

Capability development represents another critical pathway. Buildability enhancement is not solely technical; it is educational and organisational. Samarasinghe and Piri (2022) show that immersive virtual reality environments improve users' ability to detect constructability challenges, even at the student level. Experiential learning tools enhance spatial awareness and sequencing comprehension, strengthening the human capacity required to translate digital models into buildable solutions. Sustainable construction practices, therefore, depend on continuous professional development and digital literacy alongside technological investment.

Finally, integration must extend beyond individual projects to encompass policy and procurement reform. Sustainability and buildability are both influenced by institutional incentives. Procurement systems that prioritise the lowest upfront cost may discourage early contractor involvement and collaborative design processes, undermining buildability integration. Conversely, performance-based procurement models and lifecycle evaluation frameworks encourage coordination and long-term value creation. Moghayedi and Awuzie (2025) demonstrate in the context of net-zero housing delivery that systemic alignment across policy, finance, and stakeholder collaboration is essential for sustainable outcomes. Buildability integration similarly requires enabling governance environments.

Integrating buildability into sustainable construction is not a singular technical intervention but a systemic transformation. It requires early-stage embedding, socio-technical alignment, structured feedback loops, measurable evaluation frameworks, capability development, and supportive policy structures. When these pathways operate in concert, buildability becomes embedded as a proactive governance mechanism that safeguards sustainability performance across the project lifecycle.

9.3 Advanced Technologies and Innovations for Buildability Improvement

Technological transformation is reshaping the meaning and operationalisation of buildability within contemporary construction systems. Historically reliant on expert judgment and qualitative assessment, buildability is increasingly supported by digital modelling, simulation, artificial intelligence (AI), and immersive technologies. These innovations are shifting buildability from a reactive design-checking exercise toward a predictive, data-driven, and performance-oriented capability embedded within digital construction ecosystems. Technological advancement does not replace human judgment; rather, it augments decision-making and strengthens the integration of constructability within sustainable construction strategies.

A major development in buildability practice has been the adoption of immersive visualisation tools. Virtual reality (VR) environments enable stakeholders to engage with spatial configurations, sequencing constraints, and access limitations in ways that conventional 2D drawings cannot replicate. Samarasinghe and Piri (2022) show that immersive VR significantly improves users' ability to identify constructability issues, including spatial conflicts and installation challenges. This enhancement of perceptual understanding supports earlier detection of potential problems, reducing the likelihood of rework during construction. From a sustainability perspective, immersive technologies contribute to waste reduction and efficiency by strengthening pre-construction review processes and improving shared comprehension among multidisciplinary teams.

Beyond visualisation, computational simulation methods are expanding the analytical depth of buildability assessment. Zhu et al. (2024) employ Peridynamics-based modelling to predict structural instability and failure mechanisms in complex 3D-printed concrete geometries. Unlike traditional finite element methods, this approach captures discontinuities and crack propagation during printing, allowing designers to evaluate geometric feasibility before physical fabrication begins. Such predictive capacity represents a significant advancement: buildability can be assessed before materials and time are invested in physical trials, reducing waste, energy use, and economic risk.

Additive manufacturing research also formalises buildability through performance-based criteria and testing protocols. Nerella et al. (2018, 2020) developed practice-oriented criteria for 3D-printable concretes that integrate rheological behaviour, extrusion performance, layer stability, and economic viability. Their work emphasises that buildability in digital fabrication is a coupled material–process phenomenon shaped by equipment parameters, deposition rates, and time-dependent strength development. This systems-based view extends beyond printing and offers a useful analogue for broader construction contexts where material selection, sequencing, and environmental conditions interact to determine feasibility.

AI and machine learning represent a further evolution in buildability improvement. Ballal and Sher (2003) demonstrated that artificial neural networks can model buildability considerations for preliminary structural system selection by embedding expert judgment into computational decision support. More recently, Wang et al. (2026) proposed a graph neural network approach (FB-GAT) that represents façade systems as interconnected nodes and edges, enabling predictive evaluation of buildability based on component interactions and assembly complexity. By learning from project datasets, AI models can identify patterns associated with constructability challenges and support proactive risk mitigation during design.

However, technological capability alone does not guarantee effective implementation. Human–computer interaction research underscores the importance of usability, transparency, and trust in AI-driven tools. Mitrana (2021) argues that buildability assessment systems must incorporate user-centred design principles so that practitioners can understand, interpret, and trust algorithmic outputs. Without explainability and workflow compatibility, technologically sophisticated tools risk underutilisation. Advanced buildability technologies, therefore, need to be embedded within socio-technical systems that align digital capability with practitioner competence and organisational readiness.

Together, these technologies are transforming buildability into a predictive and optimisation-oriented discipline. Immersive tools enhance perception and shared understanding, simulation enables physics-informed feasibility analysis, AI supports data-driven prediction, and user-centred interfaces facilitate adoption. These developments strengthen the alignment between buildability and sustainability by reducing inefficiencies, enabling informed trade-offs, and supporting scalable innovation. The subsequent sections build on this technological foundation by examining industrialised construction, net-zero construction project delivery, and resilience planning.

9.4 Buildability in Offsite and Modular Construction for Sustainability

The increasing adoption of off-site, modular, and industrialised construction has intensified the importance of buildability as a strategic mechanism for engendering improved sustainability performance. While off-site approaches are frequently promoted for reducing waste, improving quality control, and accelerating delivery (Obi et al., 2025), these benefits depend on rigorous integration of buildability principles across design, manufacturing, logistics, and assembly. In such industrialised systems, buildability extends beyond site-based constructability to include manufacturability, transportability, interface precision, sequencing discipline, and lifecycle reliability (Tafeni et al., 2026).

Unlike conventional site-based construction, off-site systems demand high levels of dimensional coordination and tolerance management. Errors that might be absorbed or adjusted on-site can become critical failures in modular assembly. As a result, buildability becomes inseparable from design standardisation, interface management, and production–assembly integration. The sustainability gains often associated with prefabrication, reduced material waste, and shortened construction durations, are therefore dependent on disciplined and predictable buildability performance.

Additive manufacturing research provides useful parallels for understanding buildability in industrialised systems. Nerella et al. (2018) argue that buildability criteria must reflect real-world operational conditions rather than isolated laboratory metrics. In 3D concrete printing, material properties, extrusion behaviour, deposition rates, and early-age strength development interact dynamically to determine stability. This systems-based perspective mirrors off-site environments, where buildability depends on alignment between design intent, manufacturing precision, assembly sequencing, and contextual constraints.

Casagrande et al. (2020) reinforce this argument by demonstrating that buildability outcomes are highly sensitive to testing procedures and process parameters. Translating this insight into modular construction suggests that buildability must be assessed as a system-level attribute influenced by component geometry, connection detailing, assembly tolerances, and logistics sequencing. Sustainability claims linked to prefabrication, therefore, require context-specific evaluation that accounts for process sensitivity rather than assuming universal performance benefits.

Temporal dynamics further illustrate the complexity of buildability in industrialised systems. da Silveira Júnior et al. (2024) show that time gaps between printed layers influence both stability and interlayer bonding. In modular construction, analogous temporal variables arise through curing periods, storage durations, transport delays, and phased installation schedules. These time-dependent factors influence material behaviour and connection performance, affecting durability and sustainability outcomes. Buildability in industrialised contexts is therefore dynamic and must account for process continuity and environmental variability.

The sustainability implications of effective buildability integration in off-site systems are significant. Prior research indicates that innovative prefabricated construction methods can reduce waste, lower embodied carbon, and improve energy performance in affordable housing contexts (Moghayedi & Awuzie, 2023). However, these environmental gains are contingent on alignment with local supply chains, skills availability, and affordability constraints (Jayawardana et al., 2023; Li et al., 2024). Prefabrication does not inherently guarantee sustainability; rather, buildability can be understood as a mediating factor that enable industrialised systems to be practically implemented in ways that generate context-appropriate sustainability-oriented benefits.

Overall, buildability in off-site and modular construction functions as a system-level integrator linking design, manufacturing, logistics, and assembly. When effectively managed, it improves waste reduction, quality consistency, and delivery predictability; when neglected, it can negate environmental and economic gains through rework, delay, and replacement. This emphasis on repeatability, low waste, and early-stage feasibility in industrialised construction systems aligns directly with net-zero imperatives, where emissions reduction depends on preventing rework and embedding low-carbon decisions before complexity becomes locked into delivery systems.

9.5 Buildability and Net-Zero Construction Targets

The global transition toward net-zero construction is one of the most ambitious transformations in the built environment sector. Achieving net-zero targets requires reductions in embodied carbon, operational energy consumption, construction waste, and lifecycle emissions (Awuzie et al., 2024, Moghayedi & Awuzie, 2025). While material innovation, renewable energy integration, and circular economy strategies receive significant attention, the role of buildability remains underreported in the corpus of extant literature. Yet without robust constructability integration, net-zero ambitions risk becoming technically impractical, economically infeasible, or operationally compromised. Buildability supports net-zero delivery by preventing carbon-intensive rework, enabling early-stage low-carbon optimisation, and ensuring feasibility across diverse socio-economic contexts (Al-Saaddy et al., 2025).

A critical contribution is rework reduction. Rework increases embodied emissions through additional material production, transportation, labour, and waste disposal. Singhaputtangkul et al. (2011) highlight that sustainable envelope systems must be evaluated alongside constructability requirements. Where installation complexity produces defects and corrective work, environmental gains can be eroded. In digital fabrication contexts, Zhu et al. (2024) similarly show that predictive simulation can reduce failed production and unnecessary material consumption. Nerella et al. (2020) extend this argument by emphasising economic viability as part of buildability assessment; net-zero pathways that are not cost-feasible are unlikely to be widely implemented.

Buildability also enables early-stage low-carbon optimisation. Net-zero outcomes depend heavily on early decisions regarding structural systems, material selections, building forms, and construction methodologies. Mathern et al. (2022) demonstrate that multi-objective optimisation frameworks can incorporate environmental impact indicators alongside buildability constraints, enabling transparent exploration of trade-offs and identification of solutions that are simultaneously low-impact and practicable. AI-enabled buildability assessment further supports early-stage optimisation, with Ballal and Sher (2003) demonstrating neural network decision support for preliminary system selection and Wang et al. (2026) illustrating predictive façade buildability assessment through graph neural networks. When sustainability metrics are incorporated into such predictive approaches, buildability strengthens the alignment between net-zero ambition and execution reality.

Contextual feasibility is equally central to equitable net-zero transitions. Moghayedi and Awuzie (2023) show that prefabricated methods can improve sustainability performance in affordable housing delivery in Southern Africa, but that outcomes depend on affordability, supply chain readiness, and skills availability. Moghayedi and Awuzie (2025) further propose a framework for low-income net-zero energy housing delivery that integrates policy alignment, financing mechanisms, technical guidance, and stakeholder coordination. Their findings underscore that net-zero delivery is a systemic governance and implementation challenge rather than a purely technical one. Buildability is integral to this framing because energy-efficient envelopes, renewable energy systems, and passive design strategies must remain constructible, maintainable, and affordable to succeed at scale.

Finally, net-zero delivery depends on socio-technical integration. Practitioner-focused research indicates that digital tools do not automatically improve buildability; organisational culture, collaboration practices, and skills maturity mediate effectiveness (PuwakkatiyaLokuge et al., 2025; Wimalaratne et al., 2023). Digital collaboration platforms and predictive modelling can enhance transparency and shared understanding of carbon impacts, but only where teams are equipped to use them. Similarly, Mitrana (2021) emphasises that user-centred design and explainability are essential if AI-driven buildability insights are to be trusted and acted upon.

Taken together, the reviewed studies position buildability as a central enabler of net-zero delivery. By preventing rework-related emissions, enabling early-stage optimisation, and ensuring contextual feasibility, particularly in low-income settings, buildability functions as the operational mechanism through which decarbonisation ambitions remain technically viable, economically feasible, and scalable (Mathern et al., 2022; Moghayedi & Awuzie, 2025; Nerella et al., 2020). The next section extends this analysis to resilience planning, further reinforcing buildability's role as a systems-level integrator under conditions of uncertainty and disruption.

9.6 Buildability for Resilience Planning in the Built Environment

Resilience has become a central objective in built environment discourse, encompassing the capacity of infrastructure and buildings to withstand disruption, adapt to uncertainty, and maintain functional performance under stress. While resilience is often framed in terms of structural robustness, climate adaptation, or redundancy in systems, buildability contributes fundamentally by improving execution reliability, strengthening adaptive coordination, embedding learning mechanisms, and enabling anticipatory risk mitigation across project lifecycles.

Resilience depends on reliable execution. Poor constructability can introduce vulnerabilities through defects, misaligned interfaces, and premature degradation that reduce the capacity of assets to withstand environmental and operational stress. Buildability strengthens resilience by aligning design intent with construction capability. Singhaputtangkul et al. (2011) show that envelope performance depends on constructible detailing and correct installation; failures can increase vulnerability to moisture ingress and performance deterioration. In industrialised and digital fabrication contexts, predictive assessment further enhances reliability. Zhu et al. (2024) demonstrate that simulation-based evaluation can anticipate instability during printing, enabling correction prior to failure, while da Silveira Júnior et al. (2024) highlight the sensitivity of performance to process timing, illustrating that resilience in automated construction depends on disciplined operational control.

Resilience is also an organisational and socio-technical capability. Projects inevitably face change through design revisions, supply chain disruption, environmental events, or regulatory shifts. The capacity to respond depends on communication structures, feedback mechanisms, and collaborative cultures. Wimalaratne et al. (2023) conceptualise buildability in the digital age as emerging from coordination quality and shared decision-making. These characteristics are foundational to organisational resilience because integrated teams can adapt more effectively under uncertainty. Furthermore, PuwakkatiyaLokuge et al. (2025) posit that technological advancements influence buildability through socio-organisational interactions rather than tools alone, reinforcing the need for aligned skills and governance.

Organisational learning further strengthens resilience. Henderson (2013) argues that weak design–construction feedback loops contribute to recurring buildability failures in complex projects. Without structured knowledge capture and reintegration, organisations repeat mistakes and compound inefficiency and risk. Strengthening feedback loops through post-project evaluation, digital knowledge repositories, and iterative design–construction reviews institutionalises learning, reduces systemic vulnerability, and supports continuous improvement under evolving conditions.

Predictive analytics also extends resilience by enabling anticipatory risk identification. Ballal and Sher (2003) demonstrate the potential of neural networks to support early selection of buildable systems, while Wang et al. (2026) show how graph neural networks can predict façade buildability by capturing relational complexity. Such tools can support proactive risk reduction, but, as Mitrana (2021) cautions, only if they are transparent, interpretable, and aligned with practitioner workflows.

Finally, resilience planning must account for resource constraints and governance conditions, particularly in developing contexts where infrastructure deficits and climate vulnerability intersect. Moghayedi and Awuzie (2025) show that sustainable housing delivery requires integrated policy alignment, financing mechanisms, and stakeholder coordination. Buildability strengthens resilience in such contexts by ensuring that sustainability strategies are constructible within local supply chains, skills availability, and regulatory frameworks. Overly complex solutions may undermine resilience if they depend on unavailable expertise or fragile logistics networks.

Overall, buildability strengthens resilience by improving execution reliability, enabling adaptive socio-technical coordination, institutionalising learning through feedback loops, supporting anticipatory risk management, and ensuring contextual feasibility across diverse environments. In this framing, buildability functions as a governance capability that links sustainability ambition to dependable delivery under uncertainty.

9.7 Conceptual Framing and Model: Buildability as a Socio-Technical Governance Capability

To consolidate the arguments developed across Sections 9.1–9.6, this section provides an integrative conceptual framing that positions buildability as a socio-technical governance capability linking sustainability ambition to delivery performance. The reviewed literature indicates that buildability has moved beyond being a design-stage constructability check and now operates as a multi-level mechanism through which projects manage complexity, prevent waste, and enable reliable implementation of sustainability strategies. This framing clarifies what buildability is becoming in contemporary practice and how it can be leveraged systematically in pursuit of sustainability, net-zero targets, industrialised delivery, and resilience.

A central insight from recent buildability scholarship is that buildability is co-produced by technical design attributes and the social conditions of project delivery. Osuizugbo et al. (2023) demonstrate that buildability is conceptually multidimensional, spanning design, process, organisational, and technological factors. Phenomenological studies further highlight that buildability is experienced and interpreted differently across practitioners and roles, and that its value emerges through collaboration, information continuity, and early coordination rather than isolated technical compliance (Wimalaratne et al., 2023; PuwakkatiyaLokuge et al., 2025). In this view, digital technologies do not "produce" buildability independently; they enable buildability when embedded within collaborative cultures and integrated workflows. This aligns with evidence that tool effectiveness is mediated by skills, organisational readiness, and the usability of decision-support outputs (Mitrana, 2021). Buildability improvement is therefore best understood as an interaction between technological capacity and socio-organisational integration.

A second conceptual anchor is that buildability functions as a governance capability operating across project stages rather than as a one-off design assessment. The "buildability momentum" perspective suggests that decisions made early accumulate through later stages, shaping downstream delivery outcomes and sustainability performance (Wimalaratne & Kulatunga, 2025). Similarly, research on design–construction feedback loops argues that continuous information exchange improves constructability by institutionalising learning and preventing repeated inefficiencies (Henderson, 2013). These arguments imply that buildability is a life-cycle governance mechanism that structures decision-making, coordinates actors, and maintains information continuity across phases. In practical terms, buildability becomes stabilising when embedded through early contractor involvement, collaborative procurement, structured review cycles, and performance-based evaluation; when these governance structures are fragmented, sustainability outcomes become vulnerable to redesign, rework, and value engineering.

Synthesising this chapter's arguments, buildability can be conceptualised as operating through four interacting pathways that connect project inputs to sustainability outcomes: coordination, optimisation, predictability, and learning. The coordination pathway strengthens sustainability when design–construction integration reduces miscommunication, errors, and downstream waste through early-stage alignment and digitally mediated collaboration (Wimalaratne et al., 2023; PuwakkatiyaLokuge et al., 2025). The optimisation pathway strengthens sustainability when feasibility-informed decisions are made before design choices become locked in, enabling transparent trade-off management and co-optimisation of sustainability and buildability requirements (Mathern et al., 2022; Ballal & Sher, 2003; Wang et al., 2026). The predictability pathway strengthens sustainability when standardisation and process control reduce trial-and-error and improve first-time-right performance, particularly in industrialised and digitally fabricated construction project contexts where testing configuration and process parameters significantly influence outcomes (Casagrande et al., 2020; Nerella et al., 2020; Zhu et al., 2024). The learning pathway strengthens sustainability and resilience when feedback loops and knowledge reuse reduce repeated constructability failures and enhance adaptive capacity under uncertainty (Henderson, 2013).

These pathways explain why buildability consistently appears in the literature as a bridge between sustainability intent and practical implementation: it integrates coordination, optimisation, predictability, and learning into coherent delivery systems.

9.7.1 Conceptual Framework: Buildability–Sustainability Integration Framework (BSIF)

This chapter conceptualises buildability as a socio-technical governance capability that systematically converts sustainability ambition into reliable delivery performance through four interacting pathways: coordination, optimisation, predictability, and learning.

The Buildability–Sustainability Integration Framework (BSIF) proposes a structured causal logic in which project inputs, comprising design attributes, governance arrangements, digital readiness, and organisational capability, activate these pathways. Through their interaction, the pathways shape buildability performance outcomes, including reduced uncertainty, improved assembly feasibility, lower rework incidence, and enhanced schedule reliability. These performance improvements, in turn, generate sustainability outcomes across environmental (waste and embodied carbon reduction), economic (cost certainty and productivity), and social (quality, safety, and maintainability) dimensions, while also enabling net-zero transitions and strengthening resilience capacity. The framework further recognises that the magnitude and stability of these relationships are moderated by contextual factors, including project complexity, delivery fragmentation, technology maturity, and socio-economic constraints.

Within this framing, buildability is neither a purely technical attribute nor a discrete design-stage intervention; rather, it operates as a lifecycle-integrated system that aligns digital tools, collaborative governance, and construction feasibility into a coherent mechanism for translating sustainability strategy into dependable execution. By positioning buildability at the intersection of coordination, optimisation, predictability, and organisational learning, the BSIF clarifies how sustainable construction futures can be operationalised at scale rather than remaining an aspiration.

Figure 9.1 visually synthesises these relationships, illustrating how project inputs activate the four pathways, which in turn generate buildability performance outcomes and sustainability impacts under moderating contextual conditions.

9.7.2 Implications of the Conceptual Framework

This conceptual framework implies that buildability should be evaluated and strengthened not only through technical design interventions but also through

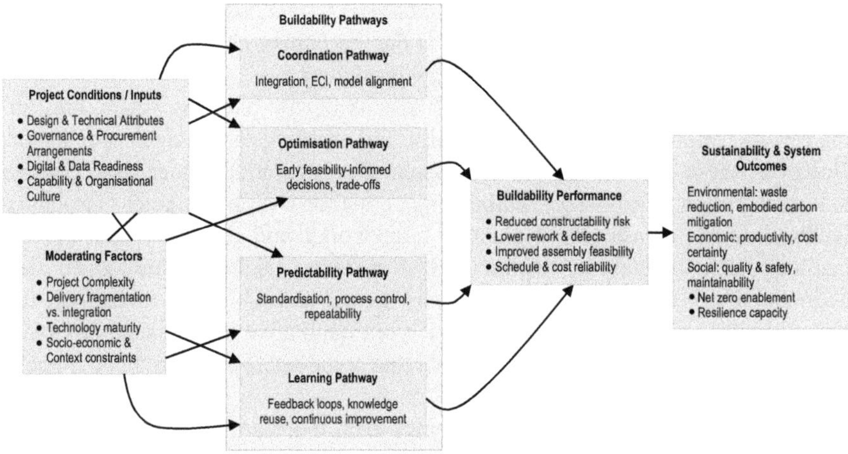

Figure 9.1 Buildability–sustainability integration framework (BSIF).

Source: Authors' Conceptualisation.

integrated governance and socio-technical capacity-building that aligns people, processes, and digital systems across the project lifecycle. For practice, the framework positions buildability as a management capability that must be institutionalised through early-stage decision gates, collaborative procurement and contractor involvement, structured design–construction feedback loops, and competence development that ensures digital tools translate into actionable constructability intelligence. This shifts buildability from a compliance-oriented check to an embedded delivery function that protects sustainability value by reducing rework, improving predictability, and strengthening lifecycle performance.

For research, the framework suggests a measurement agenda in which buildability assessment expands beyond isolated design metrics to include indicators that capture coordination quality, decision timing and lock-in effects, process predictability, and learning maturity. Such indicators would enable empirical testing of the framework's pathways and moderators, improving comparability across studies and contexts where buildability remains conceptually fragmented. This direction aligns with calls for clearer definitions, more standardised assessment methods, and stronger empirical linkage between buildability and sustainability performance (Osuizugbo et al., 2023). These implications reinforce buildability's role as a socio-technical governance capability and provide the analytical basis for this chapter's conclusion, which synthesises how buildability can operationalise sustainable construction futures.

9.8 Chapter Conclusion

This chapter has demonstrated that buildability has evolved from a narrow constructability concern into a multidimensional, socio-technical capability central to sustainable construction futures. Across sustainability performance, digital innovation, industrialised construction, net-zero transitions, and resilience planning, buildability consistently emerges as the mechanism through which environmental ambition is translated into delivery feasibility.

Buildability contributes directly to sustainability by reducing waste, preventing rework, improving lifecycle durability, and enabling early-stage optimisation. However, these benefits depend on systematic integration rather than conceptual recognition. Early-stage embedding, collaborative governance, structured feedback loops, measurable evaluation frameworks, and capability development institutionalise buildability within project systems. Digital technologies, such as immersive visualisation, predictive simulation, and AI-driven modelling, further strengthen buildability by shifting it from qualitative judgment to proactive, data-informed decision support, provided that socio-technical alignment supports effective adoption.

The importance of buildability becomes even more pronounced as the sector scales off-site and modular delivery, pursues net-zero targets, and strengthens resilience planning. Across these agendas, buildability functions as a systems-level integrator that aligns design intent, technological capability, organisational practice, and contextual feasibility into reliable delivery pathways. As sustainability imperatives accelerate globally, the future of construction will depend not only on greener materials and advanced technologies but also on coordinated and dependable delivery systems. Buildability provides the operational foundation through which sustainable construction futures can be realised in practice.

References

Al-Saaddy, A., Kabiri, D.S., & Mohareb, E. (2025, December). Embedding net-zero in construction: Client–contractor practices and collaboration. In *Proceedings of smart and sustainable built environment conference series* (pp. 395–404).

Awuzie, B., Ngowi, A., & Aghimien, D. (2024). Towards built environment decarbonisation: A review of the role of Artificial intelligence in improving energy and Materials' circularity performance. *Energy and Buildings*, 319, 114491.

Ballal, T.M., & Sher, W.D. (2003). Artificial neural network for the selection of buildable structural systems. *Engineering, Construction and Architectural Management,* 10(4), 263–271.

Casagrande, L., Esposito, L., Menna, C., Asprone, D., & Auricchio, F. (2020). Effect of testing procedures on buildability properties of 3D-printable concrete. *Construction and Building Materials,* 245, 118286.

Caroli, T., Campioli, A., & Lavagna, M. (2022, May). Reversible, sustainable and circular constructive systems: Buildability conditions. In *International Symposium: New Metropolitan Perspectives* (pp. 1860–1869). Springer International Publishing.

Crowther, P. (2002). Design for buildability and the deconstruction consequences. In *Proceedings of the 3rd annual meeting of CIB task group 39* (pp. 1–9). CIB, Int Council for Research and Innovation in Building Conn.

da Silveira Júnior, J.G., de Moura Cerqueira, K., de Araújo Moura, R.C., de Matos, P.R., Rodriguez, E.D., de Castro Pessôa, J.R., & Tramontin Souza, M. (2024). Influence of time gap on the buildability of cement mixtures designed for 3D printing. *Buildings,* 14(4), 1070.

Dasari, S.K., Trovalusci, P., Fantuzzi, N., Pingaro, M., & Panei, R. (2023, June). Sustainable spatial structures: A design approach using shape and topology optimization to minimize environmental impact and improve buildability. In *Italian workshop on shell and spatial structures* (pp. 279–288). Springer Nature Switzerland.

Diaz Schery, C.A., Caiado, R.G.G., Corseuil, E.T., & Neves, H. (2025). Bridging BIM maturity with emerging paradigms: understanding the role of industry 4.0 and sustainability in the digital transformation of construction. *Construction Innovation: Information Process Management*,. https://doi.org/10.1108/CI-10-2024-0305

Henderson, J.R. (2013). Enhancing buildability through improving design–construction feedback loops within complex projects. Doctoral dissertation, Loughborough University.

Hyde, R. (1995). Buildability as a design concept for architects: A case study of laboratory buildings. *Engineering, Construction and Architectural Management*, 2(1), 45–56.

Jayawardana, J., Kulatunga, A.K., Jayasinghe, J.A.S.C., Sandanayake, M., & Zhang, G. (2023). Environmental sustainability of off-site construction in developed and developing regions: A systematic review. *Journal of Architectural Engineering*, 29(2), 04023008.

Karimi, R., Shirowzhan, S., & Sepasgozar, S.M. (2023). Architectural design criteria considering the circular economy and buildability for smart disaster relief shelter prototyping. *Buildings*, 13(7), 1777.

Laovisutthichai, V., Lu, W., & Bao, Z. (2022). Design for construction waste minimization: Guidelines and practice. *Architectural Engineering and Design Management*, 18(3), 279–298.

Lesniewski, L., & Berkebile, B. (2020). Sustainable design and construction, integrated delivery processes, and building information modeling. In V. Loftness (Ed.), *Sustainable built environments* (pp. 551–572). Springer US. https://doi.org/10.1007/978-1-0716-0684-1_390

Li, C.Z., Wen, S., Yi, W., Wu, H., & Tam, V.W. (2024). Offsite construction supply chain challenges: An integrated overview. *Journal of Construction Engineering and Management*, 150(7), 03124003.

Love, P.E., Smith, J., Ackermann, F., Irani, Z., & Teo, P. (2018). The costs of rework: Insights from construction and opportunities for learning. *Production Planning & Control*, 29(13), 1082–1095.

Mahamid, I. (2022). Impact of rework on material waste in building construction projects. *International Journal of Construction Management*, 22(8), 1500–1507.

Mathern, A., Penadés-Plà, V., Armesto Barros, J., & Yepes, V. (2022). Practical metamodel-assisted multi-objective design optimization for improved sustainability and buildability of wind turbine foundations. *Structural and Multidisciplinary Optimization,* 65(2), 46.

Mbamali, I., Aiyetan, O.A., & Kehinde, J.O. (2005). Building design for buildability: An investigation of the current practice in Nigeria. *Building and Environment*, 40(9), 1267–1274.

Mitrana, G. (2021). User-centred design for the interface of an artificial intelligence buildability tool. Master's dissertation, ETSI Informática.

Moghayedi, A., & Awuzie, B. (2023). Towards a net-zero carbon economy: A sustainability performance assessment of innovative prefabricated construction methods for affordable housing in Southern Africa. *Sustainable Cities and Society,* 99, 104907.

Moghayedi, A., & Awuzie, B.O. (2025). A framework for facilitating low-income net-zero energy housing delivery in developing countries: Insights from a practical case in South Africa. *Building and Environment,* 276, 112847.

Nerella, V.N., Krause, M., & Mechtcherine, V. (2018). Practice-oriented buildability criteria for developing 3D-printable concretes in the context of digital construction.

Nerella, V.N., Krause, M., & Mechtcherine, V. (2020). Direct printing test for buildability of 3D-printable concrete considering economic viability. *Automation in Construction*, 109, 102986.

Obi, L.I., Awuzie B.O., Asare, O., Lamb, S, & Thurairajah, N. (2025). A circularity cost framework for offsite housing projects. *Construction Innovation: Information Process Management*,. https://doi.org/10.1108/CI-02-2025-0075

Ogbeifun, E., Zwalda, N., Imil, I.A., & Pretorius, J.H.C. (2023). Harmonising aesthetics, functionality, stability, buildability, serviceability, maintainability, and cost in infrastructure development. In M.B. Jelodar, R. Lovreglio, Z. Feng, D. Paes, V. Kumar, C. Siriwardana (Eds.), *Creating capacity and capability: Embracing advanced technologies and innovations for sustainable future in building education and practice*. AUBEA 2023. Lecture Notes in Civil Engineering (vol. 564, pp. 3–15). Springer. https://doi.org/10.1007/978-981-96-2802-5_1

Osuizugbo, I.C. (2025). Buildability-led procurement: A new model for future construction projects delivery. *Construction Innovation: Information Process Management*,. https://doi.org/10.1108/CI-10-2023-0272

Osuizugbo, I.C., Okolie, K.C., Oshodi, O.S., & Oyeyipo, O.O. (2023). Buildability in the construction industry: A systematic review. *Construction Innovation: Information Process Management*, 23(5), 1300–1322.

Pheng Low, S. (2001). Quantifying the relationships between buildability, structural quality and productivity in construction. *Structural Survey*, 19(2), 106–112.

PuwakkatiyaLokuge, I.W., Kulatunga, U., Gajendran, T., & Beard, C. (2025). Technological advancements and their influence on buildability in construction projects: A phenomenological analysis. *Construction Innovation: Information Process Management*. https://doi.org/10.1108/CI-04-2025-0153

Samarasinghe, D.A.S., & Piri, I.S. (2022). Assessing design buildability through virtual reality from the perspective of construction students. *Built Environment Project and Asset Management,* 12(5), 823–836.

Singhaputtangkul, N., Low, S.P., Teo, A.L., & Hwang, B.G. (2014). Criteria for architects and engineers to achieve sustainability and buildability in building envelope designs. *Journal of Management in Engineering*, 30(2), 236–245.

Singhaputtangkul, N., Low, S.P., & Teo, A.L. (2011). Integrating sustainability and buildability requirements in building envelopes. *Facilities,* 29(5/6), 255–267.

Tafeni, L., Awuzie, B., Aghimien, D. (2026). Exploring challenges and strategies for advancing circular economy principles in prefabricated construction in South Africa. In: Aigbavboa, C., Awuzie, B., Aghimien, D., Oke, A.E., & Omotayo, T. (Eds.) *Facilitating inclusivity in multi-, inter-, and transdisciplinary sustainable built environment research in emerging economies*. SURE-Built 2025. Lecture Notes in Civil Engineering, Vol. 772. Springer. https://doi.org/10.1007/978-3-032-08992-2_74

Wang, B., Lu, W., Wu, L., Gao, Y., Peng, Z., & Kristof, C. (2026). FB-GAT: A graph neural networks (GNNs) approach to assessing facades' buildability. *Advanced Engineering Informatics,* 69, 103898.

Wimalaratne, P.L.I., & Kulatunga, U. (2025). Buildability momentum across project stages: Empirical insights from RIBA plan of work towards sustainable construction. In *Proceedings of the 13th world construction symposium* (pp. 332–346).

Wimalaratne, P.L.I., Kulathunga, U., & Gajendran, T. (2021). Comparison between the terms constructability and buildability: A systematic literature review. In *Proceedings of the 9th world construction symposium* (pp. 196–207).

Wimalaratne, P.L.I., Kulatunga, U., & Gajendran, T. (2023). 'Buildability' in the digital age: A phenomenological discourse of industry practitioners' perceptions. *Buildings,* 13(11), 2870.

Wong, F.W.H., Lam, P.T.I., Chan, A.P.C., & Chan, E.H.W. (2006). A review of buildability performance in Hong Kong and strategies for improvement. *Surveying and Built Environment,* 17(2), 37–48.

Zhu, J., Cervera, M., & Ren, X. (2024). Buildability of complex 3D-printed concrete geometries using peridynamics. *Automation in Construction,* 165, 105575.

10 Conclusion and Future Pathways

10.1 Summary and Synthesis

This monograph has examined buildability as a central organising capability within contemporary construction and infrastructure delivery systems. Across its theoretical, empirical, and applied chapters, this book has advanced the argument that buildability should no longer be understood as a narrow technical concern associated with detailing or site logistics, but rather as a socio-technical governance capability that operates across the entire project lifecycle.

As established in Chapter 1, the motivation for this research emerged from persistent underperformance within the global construction sector despite decades of reform initiatives. Cost overruns, schedule delays, safety incidents, quality failures, and environmental inefficiencies remain systemic rather than exceptional (Love et al., 2020). Building on this diagnosis, the monograph proposed that these failures are not primarily attributable to technical incompetence, but to fragmented governance structures that separate design intent from delivery capability.

The historical analysis presented in Chapter 2 traced the intellectual foundations of buildability to early critiques of construction industry organisation, most notably the Emmerson Report (Emmerson, 1962). These critiques highlighted the structural disconnection between design and construction functions and established the foundational insight that inefficiencies are often embedded upstream in design and procurement decisions. This chapter demonstrated that the logic underpinning buildability is not new; rather, what has evolved is the degree to which it has been formalised, institutionalised, and operationalised.

The institutionalisation of buildability was examined in Chapter 3, which documented its transition from tacit professional intuition to a codified governance framework through initiatives such as CIRIA in the United Kingdom (CIRIA, 1983), the constructability programme of the Construction Industry Institute in the United States (CII, 1993), and Singapore's Buildable Design Appraisal System (BCA, 2005). This transition marked a critical shift in the governance of construction knowledge, from informal craft-based judgement toward structured, assessable, and policy-relevant systems.

Conceptual clarification was provided in Chapter 4, which examined the relationship between buildability and constructability traditions. While these concepts

DOI: 10.1201/9781003747284-14

originated in different regional and professional contexts, the analysis demonstrated their convergence within digitally enabled and collaborative project environments (Wimalaratne et al., 2021). Both traditions now operate as mechanisms for integrating construction knowledge into early-stage decision-making, and their distinction is increasingly institutional rather than functional.

The theoretical framework developed in Chapter 5 situates buildability within a composite analytical lens drawing on rationalisation and standardisation theory (Illingworth, 1984), socio-technical systems thinking, transaction cost economics, agency theory, and signalling theory. This framework establishes buildability as a coordination mechanism that reduces uncertainty, aligns incentives, and enhances predictability across fragmented delivery systems.

Empirical and applied insights were presented in Chapters 6 through 8, which examined buildability in relation to procurement systems, digitalisation, safety performance, and organisational learning. These chapters demonstrated that buildability outcomes are contingent upon governance arrangements rather than isolated technical interventions. Most significantly, Chapter 9 repositioned buildability within sustainability and net-zero transitions, arguing that environmental performance is fundamentally dependent on early-stage coordination and lifecycle governance (Love et al., 2020; Firoozi et al., 2024).

Across these chapters, a consistent proposition emerged: buildability operates as a system-level governance capability that shapes how infrastructure is conceived, coordinated, executed, and sustained. This concluding chapter synthesises these findings, articulates their implications, and identifies pathways for future research, practice, and policy development.

10.2 Conclusions

Several interrelated conclusions emerge from the analysis presented throughout this monograph. Together, they reposition buildability from a peripheral design concern to a central organising principle for construction governance in an era defined by complexity, sustainability imperatives, and institutional reform.

10.2.1 Buildability as a Lifecycle Governance Capability

The first and most foundational conclusion of this monograph is that buildability must be conceptualised as a lifecycle governance capability rather than a discrete technical intervention applied during design development. As demonstrated in Chapters 1 and 2, fragmentation has long been recognised as a defining structural characteristic of construction systems, shaping how decisions are made, responsibilities are allocated, and risks are transferred (Emmerson, 1962). Despite decades of reform initiatives, this fragmentation continues to constrain performance optimisation across the sector.

Evidence reviewed in Chapters 3 and 6 confirms that a substantial proportion of project outcomes, relating to cost certainty, programme reliability, quality performance, safety, and environmental impact, are determined during early design and planning stages (CIRIA, 1983; CII, 1993). Decisions concerning structural systems, dimensional coordination, material selection, construction sequencing, and tolerances embed consequences that are costly or impossible to reverse once construction is underway. These findings reinforce the systems-theoretical perspective developed in Chapter 5, which conceptualises infrastructure delivery as a socio-technical system characterised by interdependence, path dependency, and non-linear causality.

Within such systems, buildability functions as a governance mechanism that aligns design intent with delivery capability across the project lifecycle. It mediates the interaction between professional knowledge domains, organisational boundaries, and technical artefacts by regulating decision timing, structuring information flows, and embedding accountability for downstream consequences. In this sense, buildability does not merely improve constructability at the point of execution; it governs how complexity is managed and coordinated from project inception through delivery and into operation.

The analysis further demonstrates that treating buildability as a checklist exercise or late-stage design review, as critiqued in Chapter 7, is analytically and operationally insufficient. Such approaches focus on identifying symptoms of poor coordination rather than addressing underlying governance failures. Effective buildability integration requires institutionalised processes that ensure construction knowledge is embedded at points of maximum leverage within the lifecycle, supported by clear decision rights and performance accountability.

This monograph, therefore, concludes that buildability should be understood as a foundational governance capability embedded across project phases. Its effectiveness depends not on isolated technical competence, but on the extent to which delivery systems are structured to integrate knowledge, align incentives, and manage interdependence throughout the lifecycle of the built asset.

10.2.2 Fragmentation and the Centrality of Construction Procurement Reform

The second conclusion concerns the persistent role of fragmentation and the central importance of procurement systems in shaping buildability outcomes. As established in Chapters 2 and 6, fragmentation has long been identified as a defining structural condition of the construction sector, characterised by the institutional separation of design, construction, and production knowledge (Emmerson, 1962). Despite successive waves of reform, this fragmentation continues to undermine coordination, embed inefficiencies into early-stage decisions, and constrain the effective integration of buildability across project lifecycles.

Traditional procurement models, particularly those predicated on the sequential separation of design and construction responsibilities, institutionalise this fragmentation. As analysed in Chapter 6, such arrangements generate knowledge asymmetries between designers and constructors, delay constructability feedback, and exacerbate agency conflicts. From the perspective of transaction cost economics and agency theory outlined in Chapter 5, these procurement structures increase coordination costs and incentivise opportunistic behaviour, resulting in designs that are formally compliant but operationally inefficient (Moore, 1993).

Empirical evidence reviewed throughout the monograph demonstrates that procurement models incorporating early contractor involvement, collaborative contracting, or integrated project delivery significantly enhance buildability outcomes. By enabling construction expertise to inform design decisions at an early stage, these models reduce coordination failures, support joint optimisation of design and construction processes, and mitigate downstream disputes. As shown in Chapter 6, such arrangements consistently improve cost certainty, programme reliability, and construction safety performance (CII, 1993).

However, the analysis also demonstrates that procurement reform should not be understood as a purely contractual adjustment. Rather, it represents a form of institutional realignment. Without procurement frameworks that prioritise lifecycle value over lowest initial cost, buildability initiatives remain vulnerable to erosion through late-stage value engineering and risk transfer. When cost minimisation is privileged over constructability and long-term performance, buildability considerations are often marginalised, undermining both delivery efficiency and sustainability outcomes (Love et al., 2020).

The monograph, therefore, concludes that procurement reform is a strategic precondition for institutionalising buildability as a governance capability. Procurement systems shape not only contractual relationships but also the timing of decisions, the flow of knowledge, and the alignment of incentives across project participants. Where procurement frameworks support early collaboration and shared accountability, buildability can function as an effective coordination mechanism; where they do not, buildability remains constrained regardless of technical competence or intent.

10.2.3 Digitalisation as Socio-Technical Augmentation, Not Substitution

The third conclusion concerns the role of digital transformation in operationalising buildability across contemporary project delivery systems. As demonstrated in Chapter 7, advances in building information modelling, 4D and 5D simulation, digital twins, and computational design have substantially expanded the analytical capacity available to project teams (Tramontin et al., 2022; Wimalaratne et al., 2023). These technologies enable proactive constructability analysis, early clash detection, sequencing optimisation, and scenario testing at a level of precision that was previously unattainable within traditional document-based workflows.

This digital capability fundamentally alters the temporal logic of buildability. Rather than functioning as a reactive practice focused on resolving constructability issues after they emerge on-site, digital tools allow buildability to operate as a predictive governance capability. Design options can be evaluated against construction constraints before commitments are locked in, enabling early identification of risks related to sequencing, access, tolerances, logistics, and supply chain interfaces. In doing so, digital platforms enhance transparency, support shared understanding across disciplinary boundaries, and facilitate evidence-based decision-making during the stages of greatest leverage.

However, this monograph consistently underscores critical caution against technological determinism. As argued in Chapters 5 and 7, digital tools are socio-technical artefacts whose effectiveness is mediated by organisational culture, institutional incentives, and human capability. Without aligned governance structures, particularly those governing decision authority, accountability, and information ownership, digitalisation can reinforce fragmentation rather than resolve it. Highly sophisticated models may coexist with adversarial procurement arrangements, siloed professional responsibilities, and late-stage decision-making processes that neutralise their potential value.

Buildability therefore remains a fundamentally human-centred coordination capability. Digital technologies augment this capability by extending cognitive reach, enabling simulation, and supporting collaborative sense-making, but they do not substitute for the institutional arrangements that govern how decisions are made and enacted. The persistent underperformance of many digital initiatives within the construction sector reflects not technological limitation, but governance misalignment. Digitalisation enhances buildability only when embedded within delivery systems that reward early coordination, shared optimisation, and lifecycle accountability.

10.2.4 Sustainability, Net-Zero Transitions, and the Buildability Nexus

A fourth conclusion concerns the increasingly central role of buildability in sustainability and net-zero transitions within the built environment. As established in Chapter 9, construction constitutes a major contributor to global carbon emissions, not only through operational energy use but also through embodied carbon associated with material production, transportation, site activities, and rework (Love et al., 2020). Empirical studies consistently demonstrate that design changes, inefficient sequencing, and corrective demolition significantly inflate embodied carbon footprints, often eroding sustainability gains that were envisaged at the conceptual stage.

Buildability directly addresses these sources of environmental inefficiency. By improving predictability, reducing rework, and enabling prefabrication and modularisation, buildability lowers material waste, energy consumption, and carbon intensity across project lifecycles. More fundamentally, it embeds environmental considerations into early-stage decision-making, where leverage over both cost and carbon outcomes is greatest. In this sense, buildability functions not merely as a technical efficiency mechanism but as a governance pathway through which sustainability objectives are rendered operationally feasible.

Within the Built System Integration Framework (BSIF) developed in Chapter 9, sustainability is therefore conceptualised not as an isolated or competing objective, but as an emergent property of governance quality. Projects characterised by strong buildability governance tend to demonstrate superior environmental performance because coordination mechanisms align sustainability ambitions with technical feasibility, supply chain capability, and economic viability. Conversely, weak buildability integration often results in sustainability measures being treated as discretionary, leaving them vulnerable to compromise under cost and programme pressures (Firoozi et al., 2024).

Circular economy strategies further reinforce this relationship. Approaches such as design for disassembly, modular construction, prefabrication, and material standardisation are inherently buildability-dependent. They require disciplined dimensional coordination, standardised interfaces, and integrated supply chains to function effectively. Without robust buildability governance, circularity remains aspirational rather than operational, confined to policy rhetoric rather than realised performance.

Net-zero commitments are particularly exposed to governance failure. When environmental objectives are not embedded within constructible design logic, they are frequently sacrificed during late-stage value engineering as projects confront budgetary and scheduling constraints. Buildability mitigates this risk by embedding technical realism into conceptual and schematic design, ensuring that sustainability goals are feasible, costed, and constructible from the outset. In doing so, buildability acts as a stabilising mechanism that protects environmental ambition from downstream erosion and translates net-zero commitments into deliverable outcomes.

10.2.5 Buildability, Equity, and Inclusive Development

The fifth conclusion extends the analysis beyond efficiency and environmental sustainability to questions of equity, public value, and inclusive development. As discussed in Chapters 1 and 8, infrastructure deficits in rapidly urbanising regions and developing economies are frequently compounded by governance weaknesses rather than purely financial or technical constraints. Poorly coordinated and overly complex designs often result in cost escalation, extended delivery periods, elevated maintenance burdens, and premature asset failure, outcomes that disproportionately affect low-income communities and resource-constrained public authorities (Dosumu & Aigbavboa, 2018).

Buildability offers a critical pathway toward more equitable infrastructure delivery by promoting simplicity, durability, and maintainability within design and delivery systems. Designs that are demonstrably buildable are more likely to be delivered on time and within budget, thereby reducing fiscal pressure on public clients and end-users. More importantly, buildable designs are typically aligned with available construction skills, supply chains, and maintenance capacities, increasing the likelihood that assets can be operated and maintained effectively over their intended lifecycles. As shown in Chapter 8, this alignment is particularly significant in contexts where long-term operational capacity is limited and reliance on specialised technologies or imported expertise is unsustainable.

From a governance perspective, buildability is therefore not value-neutral. It embodies normative commitments to reliability, affordability, and stewardship of public resources. By embedding constructability and lifecycle considerations into early-stage decision-making, buildability reduces the risk that infrastructure investments generate hidden social costs through failure, underperformance, or premature replacement. In this sense, buildability contributes to the realisation of public value by ensuring that infrastructure performs as intended for its users over time, rather than merely achieving initial delivery targets.

Integrating buildability principles into public housing, healthcare, education, and community infrastructure programmes thus carries significant equity implications. When buildability is institutionalised within public-sector governance frameworks, infrastructure is more likely to support inclusive access, long-term serviceability, and intergenerational resilience. As the analysis in Chapter 8 demonstrates, buildability strengthens the capacity of infrastructure systems to deliver social outcomes reliably, reinforcing its role not only as a technical or managerial concern but also as a foundational component of equitable and responsible development.

10.3 Implications for Research, Practice, and Policy

10.3.1 Research Implications

The analysis presented in this monograph identifies several priorities for future research aimed at consolidating buildability as a rigorous and operationalisable governance concept. While existing studies have established strong conceptual and qualitative links between buildability and project performance, the findings of Chapters 5–9 indicate that further empirical validation is required to strengthen causal explanation and comparative generalisation.

A primary research priority concerns the development of standardised and quantifiable buildability performance metrics. As discussed in Chapters 3 and 5, current approaches to buildability assessment remain fragmented, often focusing on isolated design attributes or compliance-based scoring systems. Future research should aim to develop integrative buildability indices that capture cost, time, quality, safety, and sustainability outcomes across the project lifecycle. Such indices would enable cross-project and cross-regional comparison, support longitudinal analysis, and enhance the policy relevance of buildability research.

Longitudinal and lifecycle-oriented studies represent a second critical research direction. As demonstrated in Chapters 6 and 9, the effects of early-stage buildability decisions frequently materialise during construction and operation, yet most empirical studies remain confined to single project phases. Longitudinal research designs tracking projects from conceptual design through delivery and into operation would provide greater insight into causal mechanisms, feedback loops, and path dependencies. This approach would also support a more robust assessment of how buildability influences asset durability, maintenance performance, and lifecycle carbon outcomes.

A third research frontier lies in the intersection between buildability and emerging digital technologies. While Chapter 7 demonstrated the potential of BIM, digital twins, and advanced simulation tools to enhance constructability analysis, there remains a limited understanding of how these technologies interact with governance structures and institutional contexts. Future research should therefore examine AI-enabled constructability modelling, automated design optimisation, and digital twin governance through a socio-technical lens, focusing on decision authority, accountability, and organisational capability rather than technical performance alone.

Comparative and institutional research constitutes a further priority. As highlighted in Chapters 3 and 8, buildability practices are shaped by regulatory environments, procurement regimes, and professional cultures. Cross-national and cross-sectoral comparative studies would deepen understanding of how institutional arrangements enable or constrain buildability integration, and why similar technical tools produce divergent outcomes across contexts. Such research would strengthen theoretical coherence while enhancing the transferability of findings.

Finally, the monograph points to the need for greater theoretical integration between buildability research and broader bodies of knowledge in governance, institutional economics, and public value theory. As argued in Chapter 5, buildability operates as a coordination and governance capability rather than a purely technical attribute. Future research that explicitly situates buildability within these theoretical traditions would advance conceptual clarity and position buildability scholarship more firmly within mainstream infrastructure and project governance literature.

10.3.2 Practice Implications

The findings of this monograph carry significant implications for professional practice across the construction and infrastructure sectors. As demonstrated throughout Chapters 6–8, buildability outcomes are shaped less by isolated technical competence than by how organisations structure decision-making, integrate knowledge, and manage coordination across project lifecycles. Practice-oriented reform must therefore move beyond ad hoc interventions toward institutionalised processes that embed buildability as a routine governance function.

A central implication for practitioners is the need to formalise buildability integration at defined decision gateways. As shown in Chapter 6, reliance on informal reviews or individual experience is insufficient in complex, multi-actor delivery environments. Structured buildability reviews, conducted at conceptual, schematic, and detailed design stages, enable early identification of constructability risks and prevent the downstream accumulation of inefficiencies. These reviews must be multidisciplinary in nature, incorporating design, construction, and supply chain perspectives, and supported by clear accountability for acting on identified issues.

Early contractor involvement emerges as a second critical practice implication. Evidence presented in Chapters 6 and 7 demonstrates that integrating construction expertise during early design stages significantly improves coordination quality, cost certainty, and programme reliability. However, early involvement alone is insufficient if not accompanied by genuine influence over design decisions. Practitioners must therefore ensure that collaborative arrangements are supported by governance structures that allocate decision authority appropriately and align incentives with joint optimisation rather than risk transfer.

Digital tools represent a further area of practical significance, but their deployment requires careful alignment with organisational capability. As analysed in Chapter 7, building information modelling, 4D and 5D simulation, and digital twin platforms can substantially enhance buildability analysis when used to support proactive decision-making. In practice, this requires investment in digital literacy, clear protocols for information exchange, and integration of digital outputs into formal governance processes. Without such alignment, digital tools risk becoming visualisation artefacts rather than decision-support systems.

The monograph also highlights the importance of organisational learning and knowledge retention. As discussed in Chapter 8, many buildability failures recur because constructability insights are not systematically captured or transferred across projects. Practitioners should strengthen post-project review processes, knowledge repositories, and feedback mechanisms to ensure that lessons learned inform future decision-making. Embedding buildability knowledge within organisational memory is essential for achieving cumulative performance improvement rather than isolated success.

Finally, the analysis underscores the need for cultural change within professional practice. Buildability thrives in environments characterised by trust, collaboration, and shared accountability. Adversarial behaviours, siloed responsibilities, and short-term optimisation undermine its effectiveness. As this monograph has demonstrated, effective buildability integration requires practitioners to adopt a lifecycle-oriented mindset, recognising that early coordination and shared problem-solving are central to delivering reliable, sustainable, and equitable outcomes.

10.3.3 Policy Implications

The findings reported in this monograph have significant implications for public policy and regulatory frameworks governing the delivery of the built environment. As demonstrated across Chapters 2, 3, 6, and 9, buildability outcomes are strongly shaped by institutional settings that influence procurement practices, decision timing, accountability structures, and performance incentives. Policy interventions therefore play a critical role in either enabling or constraining the integration of buildability as a lifecycle governance capability.

A primary policy implication concerns the alignment of regulatory and procurement frameworks with lifecycle value creation. As analysed in Chapters 3 and 6, policy regimes that prioritise the lowest initial cost tend to externalise downstream consequences, undermining buildability integration and incentivising short-term optimisation. Governments and public clients can counteract this tendency by embedding buildability and lifecycle performance criteria within procurement policies, funding mechanisms, and evaluation frameworks. Such alignment encourages early coordination, constructability-informed design, and shared responsibility for long-term outcomes.

The use of buildability appraisal and assessment systems represents a second area of policy relevance. As discussed in Chapter 3, instruments such as buildability scoring mechanisms demonstrate how regulatory tools can influence industry behaviour by signalling expectations and rewarding coordination quality. When carefully designed, such systems can promote standardisation, prefabrication, and dimensional coordination without prescribing specific technical solutions. However, the analysis cautions that appraisal systems must be integrated within broader governance frameworks to avoid devolving into compliance-driven exercises that prioritise scoring over substantive performance.

Policy also has a central role to play in supporting sustainability and net-zero transitions. As established in Chapter 9, environmental objectives are most effectively achieved when embedded within early-stage decision-making processes rather than retrofitted through late-stage regulation. Policy frameworks that integrate buildability with sustainability targets, particularly embodied carbon reduction and circular economy principles, can enhance delivery feasibility while safeguarding environmental ambition. In this context, buildability functions as a policy enabler that translates aspirational targets into constructible outcomes.

Capacity building within public-sector institutions constitutes a further policy priority. As highlighted in Chapters 6 and 8, public clients often lack the organisational capability required to specify, evaluate, and govern buildability effectively. Investment in professional development, guidance frameworks, and institutional learning mechanisms is therefore essential. Without such capacity, even well-designed policies risk underperforming due to inconsistent implementation or reliance on external consultants.

Finally, the monograph underscores the importance of policy coherence and stability. Frequent regulatory change, fragmented oversight, and misaligned policy objectives can undermine long-term investment in buildability capability. Policymakers should therefore view buildability not as a discrete regulatory requirement, but as a foundational principle underpinning infrastructure governance, sustainability strategy, and public value creation.

In summary, effective policy intervention does not mandate buildability through prescriptive regulation alone. Rather, it creates the institutional conditions, through procurement reform, performance frameworks, capacity building, and sustainability alignment, within which buildability can function as a systemic governance capability. Where such conditions are present, buildability can contribute meaningfully to reliable, sustainable, and equitable infrastructure delivery.

10.4 Closing Reflections and Future Pathways

Buildability has evolved from a peripheral design consideration into a foundational socio-technical governance capability. Historically rooted in critiques of fragmentation, it now operates at the intersection of digital transformation, sustainability imperatives, and institutional reform within the built environment.

The contemporary construction sector faces unprecedented challenges: accelerating climate urgency, rapid urbanisation, technological change, and widening socio-economic disparities. Under these conditions, constructing efficiently is no longer sufficient. Infrastructure must be delivered predictably, sustainably, and equitably across its full lifecycle if it is to meet societal expectations and long-term needs.

Buildability provides the coordination architecture through which aspiration can be aligned with execution. Sustainability commitments remain fragile when they are not embedded within dependable delivery systems. Digital tools fail to realise their potential when they are not integrated within accountable governance frameworks. Equity objectives are undermined when infrastructure is not designed for durability, maintainability, and long-term use.

The transformation of construction will not be achieved through isolated technological innovation or incremental procedural reform. Rather, it will emerge from integrated governance systems capable of disciplining complexity, aligning incentives, and translating ambition into reliable outcomes. Buildability stands at the centre of this transformation, not merely as a technical discipline, but as a normative commitment to responsible design, collaborative execution, and lifecycle accountability.

Bibliography

Building and Construction Authority (BCA). (2005). *Buildable design appraisal system: Code of practice*. BCA.

Chen, S.E., McGeorge, D., & Ostwald, M. (1993). A framework for constructability improvement. *Construction Management and Economics*, 11(2), 121–134.

CIRIA. (1983). *Buildability: An assessment*. Construction Industry Research and Information Association.

Construction Industry Institute (CII). (1986). *Constructability: A primer*. University of Texas at Austin.

Construction Industry Institute (CII). (1993). *Preview of constructability implementation*. University of Texas at Austin.

Dosumu, O., & Aigbavboa, C. (2018). *Sustainable design and construction in Africa: A system dynamics approach*. Routledge. https://doi.org/10.1201/9781351212205

Emmerson, H. (1962). *Survey of problems before the construction industries*. HMSO.

Firoozi, A.A., Firoozi, A.A., Oyejobi, D.O., Avudaiappan, S., & Flores, E.S. (2024). Emerging trends in sustainable building materials: Technological innovations, enhanced performance, and future directions. *Results in Engineering*, 24, 103521.

Grabowski, Z.J., Matsler, A.M., Thiel, C., McPhillips, L., Hum, R., Bradshaw, A., Miller, T., & Redman, C. (2017). Infrastructures as socio-eco-technical systems: Five considerations for interdisciplinary dialogue. *Journal of Infrastructure Systems*, 23(4), 02517002.

Illingworth, J.R. (1984). Buildability—Tomorrow's need. *Building Technology and Management*, 16–19.

Mohd Nawi, M.N., Baluch, N., & Bahaudin, A.Y. (2014). Impact of fragmentation issue in construction industry. *MATEC Web of Conferences*, 15.

Moore, D. (1993). Constructability and knowledge transfer. *Construction Management and Economics*, 11(4), 259–268.

Osuizugbo, I.C. (2020). The need for and benefit of buildability analysis. *Journal of Engineering, Design and Technology*, 19(5), 1207–1230.

Osuizugbo, I.C. (2025). Buildability-led procurement: A new model for future construction projects delivery. *Construction Innovation: Information Process Management*. https://doi.org/10.1108/CI-10-2023-0272

Osuizugbo, I.C., Ezeokoli, F.O., Okolie, K.C., & Olojo, A.D. (2025). Towards a successful implementation of buildability assessment in construction industry: A conceptual framework. *Journal of Engineering, Design and Technology*, 23(5), 1305–1325.

Waqar, A., Houda, M., Khan, A.M., Qureshi, A.H., & Elmazi, G. (2024). Sustainable leadership practices in construction: Building a resilient society. *Environmental Challenges*, 14, 100841.

Wimalaratne, P.L.I., Kulathunga, U., & Gajendran, T. (2021). Comparison between the terms constructability and buildability: A systematic literature review. In *Proceedings of the 9th World Construction Symposium* (pp. 196–207).

Wimalaratne, P.L.I., Kulatunga, U., & Gajendran, T. (2023). Buildability in the digital age. *Buildings*, 13(11), 2870.

Wong, F.W., Lam, P.T., Chan, E.H., & Wong, F.K. (2006). Factors affecting buildability of building designs. *Canadian Journal of Civil Engineering*, 33(7), 795–806.

Abbreviations

3D	3-Dimensioning
4D	4-Dimensioning
ADR	Alternative Dispute Resolution
BIM	Building Information Modelling
BLP	Buildability-Led Procurement
BSIF	Buildability–Sustainability Integration Framework
CDM	Construction, Design and Management
CII	Construction Industry Institute
CIRIA	Construction Industry Research and Information Association
CPD	Continuous Professional Development
DfS	Design for Safety
ECI	Early Contractor Involvement
HSE	Health and Safety Executive
HVAC	Heating, Ventilation, and Air Conditioning
ILO	International Labour Organization
IPD	Integrated project delivery
MMC	Modern Methods of Construction
OHS	Occupational Health and Safety
PBL	Project-Based Learning
PPE	Personal Protective Equipment
PtD	Prevention through Design
VR	Virtual Reality

Index

Note: **Bold** page numbers refer to tables and *italic* page numbers refer to figures.

For Product Safety Concerns and Information please contact our EU
representative GPSR@taylorandfrancis.com
Taylor & Francis Verlag GmbH, Kaufingerstraße 24, 80331 München, Germany

www.ingramcontent.com/pod-product-compliance
Ingram Content Group UK Ltd.
Pitfield, Milton Keynes, MK11 3LW, UK
UKHW020014200926
475702UK00014B/1884
* 9 7 8 1 0 4 1 2 3 6 6 2 7 *